Exploring

with the

TI-82

Graphics Calculator

Dr. Brendan Kelly

Brendan Kelly Publishing Inc.
2122 Highview Drive
Burlington, Ontario L7R 3X4

ISBN 1-895997-02-X

INTRODUCTION

The widespread acceptance of the graphing calculator as an important tool for the teaching and learning of mathematics is now evolving into a compelling assertion of its necessity. State and provincial assessment instruments are shifting from a calculator-passive position (i.e. "graphing calculators are permitted.") to a calculator-active stance (i.e. "graphing calculators are required."). Mathematics educators recognize that this new technology, though not a panacea, is an essential mathematical tool which is changing irrevocably the content and pedagogy of high school and college mathematics.

Since the graphing calculator offers a wide range of modes, functions, menus and submenus, it is not feasible to expect students to master all its functions before applying it to the study of mathematics. On the contrary, it is important that instruction in mathematics draw upon the calculator functions as needed, neglecting those aspects which are not fundamental to the mathematical development. Achieving this integration of machine and mathematics requires a sequence of carefully designed instructional activities which showcase a technology-conscious approach to each mathematical concept. This book is designed to be such a resource.

The book is divided into two parts. Part I develops the concepts of classical statistics from measures of central tendency and dispersion to correlation, trend analysis and regression. Emphasis is placed on the critical analysis of statistical inferences and an understanding of the scope and limits of statistical techniques. Part II begins with a brief introduction to the founders of probability theory. It develops the basic combinatorics required to calculate theoretical probabilities and moves quickly on to Monte Carlo methods involving modern simulation techniques for calculating empirical probabilities

Your TI-82 calculator and this book will provide for you an exciting opportunity to discover many of the important concepts in statistics and probability. You will investigate trends in the ages of academy award winning actors and actresses. You will analyse whether women will outrun men in the Olympics of the 21st Century and you will simulate coin tossing and random walk experiments. We encourage you to invest time completing the exercises and investigations, checking your solutions against those in the back of the book and reflecting upon the concepts you encounter. When you have completed the explorations in this way, you will be in command of many of the most important ideas in the mathematics of large numbers and chance. In your journey toward that destination, we hope you enjoy the content, the cartoons and the entire experience!

We believe that the teaching of mathematics to the next generation is a calling of the highest order. As you incorporate the graphing calculator into your mathematics classes, you will be introducing your students to the leading edge of instructional innovation. This book and the series of which it is a part have been designed to support your professional efforts and to facilitate your task. Please share with us your comments and suggestions for future publications.

INTRODUCTION

What's this I hear about a trilogy?

The overwhelming response from high school and college teachers to our earlier published trilogy for the TI-81 Graphics Calculator provided the inspiration for a new series for the TI-82. This book is the second of a trilogy of books of the following titles:

> •*Exploring Functions with the TI-82 Graphics Calculator*
> •*Exploring Statistics with the TI-82 Graphics Calculator*
> •*Programming and Programs for the TI-82 Graphics Calculator*

Teachers and students told us that they liked the micro-biographies of mathematicians and the historical connections underpinning the instruction. We have attempted to retain these historical perspectives in this new series and have expanded their scope. Also included is an offering of the ever delightful cartoons of Sidney Harris.

The features which these books employ to achieve the desired integration and to facilitate learning include the following:
> •worked examples with keying sequences to provide step-by-step procedures for carrying out various mathematical operations on the TI-82.
> •screen displays in both the worked examples and the solutions to the exercises, to enable students to see what should appear on their calculator screen at each stage.
> •parenthetical comments to help students interpret both the advantages and the limitations of their graphing calculator outputs.
> •detailed step-by-step solutions, rather than just answers, to the exercises.

ACKNOWLEDGEMENTS

This book attempts to teach mathematics and the use of the TI-82 Graphics calculator at the same time. To do this effectively, it has been necessary to simulate the calculator keys and fonts on the TI-82 and to generate screen displays. Achieving these elements required early access to beta versions of the TI-82 and the TI-82 Link. We are indebted to Len Catleugh and Tom Ferrio of Texas Instruments whose continuing help and support in providing early versions of software and hardware have made these publications possible.

Addison-Wesley Canada under the leadership of Tony Vander Woude has generously lent their support to this project. The knowledge and expertise of Anthony Leung, Melanie Pequeux, Linda Allison and Roberta Dick have significantly enhanced the quality of this final product.

The covers of the three books in this trilogy are the creation of Gord Pronk of *Pronk&Associates*. Study the covers carefully and you will discover why many people believe Mr. Pronk to be the most talented book designer on this continent.

For a resource to be truly effective in the classroom, it must be free of ambiguities. Instruction must be clear and activities must be sequenced so that students can learn easily from it. The only way to achieve this is to conduct extensive field testing. We are indebted to the students enrolled during the 1993-94 academic year at the Faculty of Education, University of Toronto, for working through draft versions of the manuscript and providing innumerable improvements.

A special debt of gratitude is owed to my wife, Teresa, who scrutinized the manuscript for errors that I could never detect myself. For example, she insisted that we spell Kramer vs. Kramer (see p17) with "K" in spite of my assertions that it should be spelled with a "C" in honor of the discoverer of Cramer's rule.

TABLE OF CONTENTS

©1989 By Sidney Harris - Einstein Simplified
Rutgers University Press, New Brunswick, NJ, USA

PART I — STATISTICS

TABLE OF CONTENTS

PART II — PROBABILITY

©1985 By Sidney Harris– Science Goes to the Dogs
I.S.I. Press, Philadelphia, PA, U.S.A.

The art and science of drawing conclusions from data is an intellectual minefield. So often well-intentioned people, including media reporters and researchers, collect and analyze data carefully and then draw incorrect *statistical inferences*. Sometimes information is deliberately presented in a fashion which leads others to make false inferences. For this reason, the ability to draw correct statistical inferences, has become an important literacy skill in the information age. Over 50 years ago, the great American philosopher and mathematician, Alfred North Whitehead, observed,

> *There is no more common error than to assume that, because prolonged and accurate mathematical calculations have been made, the application of the result to some fact of nature is absolutely certain.*

Before we begin to use the processing power of the graphics calculator, we must become aware of some of the pitfalls and insidious traps that lie beneath the surface of statistical data.

Analyze examples A through E and identify each error in statistical inference. What reasonable inference (if any) can be drawn from the given information in each case?

A

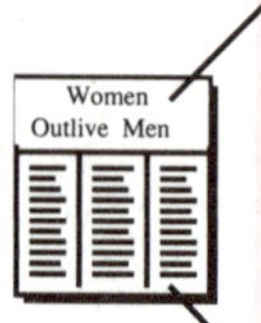

SAFER TO DRINK & DRIVE

Recently released statistics from the police files reveal that 40% of all traffic accidents are caused by drunk drivers. This implies that sober drivers are responsible for most traffic accidents. These data suggest that insurance companies should give safe driver discounts to heavy drinkers.

Some Questions to Ponder

• How is *traffic accident* defined?

• How is "cause" allocated in cases where there is shared fault?

• What percentage of all drivers on the road at any particular time are drunk?

B

Women Outlive Men

"Old age is a women's world", says Myrtle Wrinkles, a famous gerontologist. Using her own study statistics, Dr. Wrinkles told her audience that more than 50% of women over 65 years of age are widows—clear evidence that women outlive men.

Some Questions to Ponder

• How do the ages of women compare with the ages of their husbands?

• Is there a difference between the longevity of those who marry and those who do not?

• If we observed that 90% of men over 100 years of age were widowers, could we conclude men live longer than women?

C

McDougall, Littell © 1991 Reprinted with Permission

Some Questions to Ponder

•What are the four possible *rain— no rain* outcomes for the two days on the weekend?

•Is the likelihood of rain on Sunday linked to the occurrence of rain on Saturday?

D

McDougall, Littell © 1991 Reprinted with Permission

Some Questions to Ponder

•What constitutes a *user of salt?*

•What percent of the *non-users* of salt die of cancer?

E A futurist in a popular magazine, drew an inference similar to the one presented below.

In 1970 there was an average of 6 people in each car.
In 1980 there was an average of 3 people in each car.
In 1990 there was an average of 1.5 people in each car.

Conclusion: By the year 2000, every fourth car will have nobody in it!

Some Questions to Ponder

•What kind of mathematical relationship does the data suggest exists between the year and the average number of people per car?

•Why might we expect this relationship to break down by the year 2000?

Are government lotteries a tax on ignorance?

McDougall, Littell © 1991 Reprinted with Permission

7

TWO INVESTIGATIONS

In the following exercises, you are encouraged to use your calculator to perform computations.

1. The table presents real data showing the death rates from tuberculosis in 1910. These rates are given for whites and non-whites for two cities, Richmond and New York.

To determine which of these cities had the greater incidence of tuberculosis, scientists defined the *death rate from tuberculosis* as: ——————

	Population		Death from Tuberculosis	
	New York	Richmond	New York	Richmond
White	4 675 174	80 895	8365	131
Non-white	91 709	46 733	513	155
Totals	**4 766 883**	**127 628**	**8878**	**286**

$$\text{death rate from tuberculosis} = \frac{\text{number of deaths from tuberculosis}}{\text{population}}$$

a) What was the death rate from tuberculosis for whites in New York?

b) What was the death rate from tuberculosis for whites in Richmond?

c) Which city, New York or Richmond, had the higher death rate for whites?

d) Which city, New York or Richmond, had the higher death rate for non-whites?

e) What were the death rates from tuberculosis for each of the cities, New York and Richmond?

f) Which city, New York or Richmond, had the higher death rate from tuberculosis?

g) Based on your answers to parts c, d and f, which city would you, as a public health official report to have the higher incidence of tuberculosis? Explain your answer.

2. During a routine test for the HIV virus, a famous baseball player learned that he tested positive. His physician indicated that the test was 99% accurate. That is, 99% of those who have the virus test positive and 99% of those who do not have the virus test negative.

Overwhelmed with horror, the athlete rushed to his clergyman seeking spiritual guidance. The clergyman, who was also a mathematician, explained that only 0.5% of the population has the HIV virus and assured him that it was more likely that he did *not* have the virus. To convince his skeptical friend, the clergyman asked him the series of questions shown below. Upon performing the necessary calculations, the young man smiled, shook the clergyman's hand and went home to review his computations.

With the help of your calculator, complete the table. Use the table to answer these questions and decide whether the clergyman's assurances were valid.

Out of 1,000,000 people

	Number w/o Virus	Number with Virus
Number Who Test Positive		
Number Who Test Negative		
Total		

a) On average, how many people in a randomly chosen population of 1,000,000 people will have the HIV virus and how many will *not* have the virus?

b) How many of the people in part (a) will test positive?
(Remember to consider those who do *not* have the virus and yet test positive.)

c) What fraction of those people who test positive do *not* have the virus?

d) What fraction of those people who test positive do have the virus? How do you reconcile this result with the statement that the test is 99% accurate?

•Use the fact that 0.5% have the virus to complete the column totals in the table.

•Use the 99% accuracy of the test to complete the table.

•Look at the numbers in the first row to determine what percent of those who test positive have the virus.

The exercises in this lesson have been designed to encourage you to think carefully when you make inferences from data. In the remaining explorations, we will study statistical techniques which are useful but which must be applied with care.

Males OutPerform Females on Mathematics SAT

1990 PROFILE OF MATH SAT SCORES FOR MALES

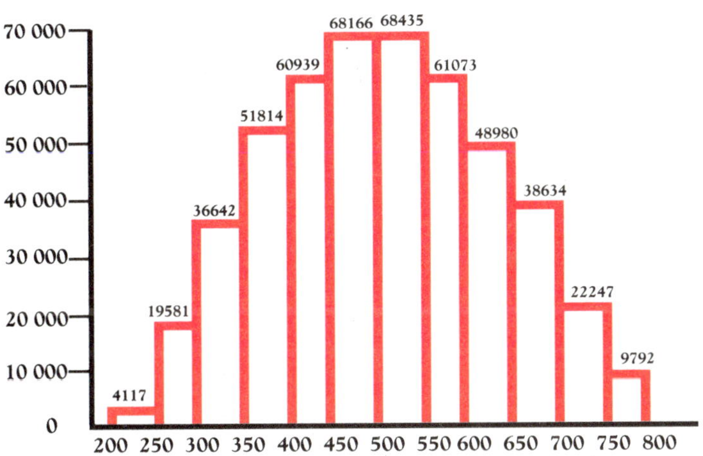

1990 PROFILE OF MATH SAT SCORES FOR FEMALES

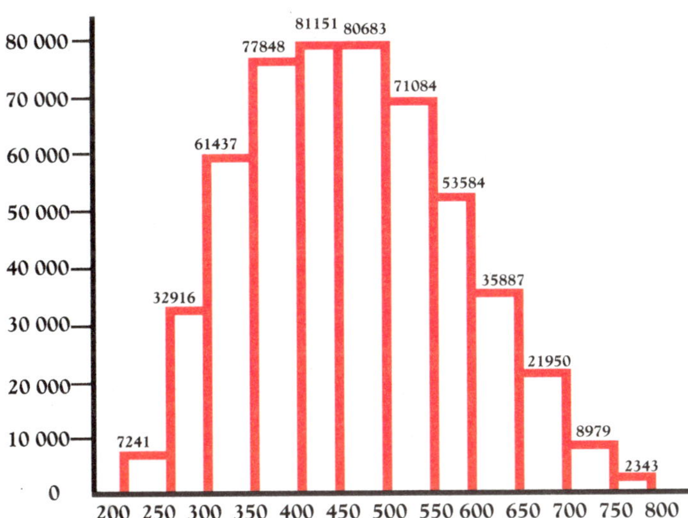

Entrance to American Colleges and Universities is determined by performance on such tests as the SAT (Scholastic Achievement Tests). The following table shows the number of males and females in 1990 whose scores on the math SAT fell within each of the given intervals. For example, the first row of the table shows that there were 4117 males and 7241 females who scored between 200 and 250 (including 200, but not including 250) on the Math SAT in 1990.

Interval	Number of Students	
	Male	**Female**
$200 \leq x < 250$	4117	7241
$250 \leq x < 300$	19581	32916
$300 \leq x < 350$	36642	61437
$350 \leq x < 400$	51814	77848
$400 \leq x < 450$	60939	81151
$450 \leq x < 500$	68166	80683
$500 \leq x < 550$	68435	71084
$550 \leq x < 600$	61073	53584
$600 \leq x < 650$	48980	35887
$650 \leq x < 700$	38634	21950
$700 \leq x < 750$	22247	8979
$750 \leq x < 800$	9792	2343
Total	**490,420**	**535,103**

Source: College Board 1991

The information in the table above is displayed for males and for females in the *histograms* on the left. The *histogram* above is a graph which shows the number of males in 1990 whose scores on the Math SAT fell within each interval given in the table. For example, we see that the vertical bar representing SAT scores in the interval given by $400 \leq x < 450$, has a height of 60939, indicating that there were 60,939 male students with scores in this interval. The corresponding histogram for female students is shown opposite.

When we compare the two histograms, we observe that the male scores seem to be distributed symmetrically about the line corresponding to $x = 500$. The female scores appear to be shifted to the left slightly, indicating a greater proportion of females below the score of 500. In interpreting these results, we must be careful not to draw quick inferences about the relative aptitudes of males and females in mathematics. Some groups have charged that the SAT tests are *biased* against females. In 1989, a New York court disallowed the awarding of scholarships based entirely on SAT scores. The College Board has announced plans to replace the SAT with a test that is *unbiased* relative to gender, race and cultural background.

WORKED EXAMPLE 1

a) Construct separately and together the histograms shown on page 9.

b) Construct the histogram of the SAT scores for males using class interval size 100. Trace along the histogram to determine the number of males who scored between 500 and 600 (including 500 but not 600).

Solution

a) The **first step** in constructing any data graph on the TI-82 is to form a table of values. To access a blank table, press: STAT ENTER .

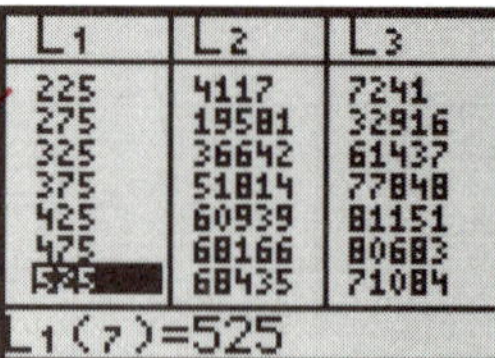

The x values in the table will be the SAT scores at the mid point of each class interval. That is, for the class interval, $200 \le x \le 250$, we enter the value 225. We define the set of all such midpoints as the list,

$$L_1 = \{225, 275, 325, 375, 425, 475, 525, 575, 625, 675, 725, 775\}$$

To enter these values in the L_1 column of the table, we position the cursor below L_1 using the arrow keys. We then type 225 and press: ENTER . We continue in this way until all the elements of L_1 have been entered.

Proceeding as above, we enter the elements of L_2 and L_3, the frequencies of SAT scores in each interval for males and for females respectively. (These are taken from the table on page 9.)

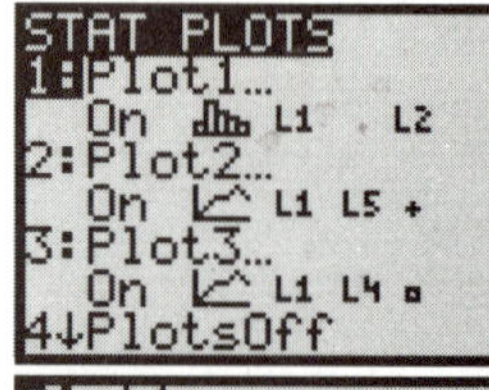

The **second step** in constructing a data graph is to define the statistical plot we want. To do this, we press: 2nd [STAT PLOT]. We obtain a display like the one shown here.

To define Plot 1 as the histogram showing the SAT scores for males, press: ENTER and select the options which are shown highlighted in the display on the right. Observe that the histogram is the rightmost icon in the line titled "Type". L_1 is the x-list and L_2 is the frequency list.

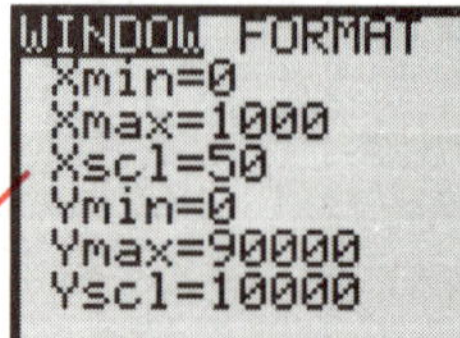

The **third step** in constructing a data graph is to define the viewing window. Press: WINDOW and then set the values as shown in the display on the right. Then we press: GRAPH to obtain the histogram for SAT scores of males, below left.

To create the corresponding histogram for SAT scores of females, we proceed as above defining Plot 2 so that L_1 is the x-list and L_3 is the frequency list. Set Plot 1 to "Off" and press: GRAPH to obtain the histogram shown in the middle. Set Plots 1 and 2 to "On" and we obtain the superimposed histograms shown below right.

Note: Xscl defines the size of the class intervals. Its value must be chosen to yield fewer than 48 intervals. Failure to set Xscl appropriately will result in an error message!

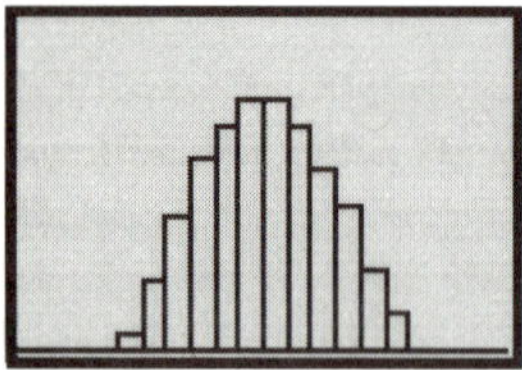
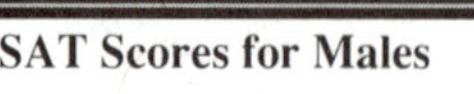

SAT Scores for Males

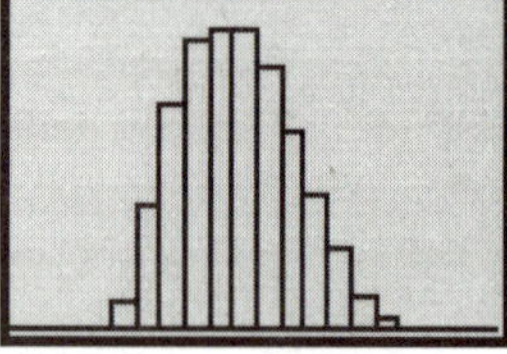

SAT Scores for Females

Superimposed Histograms

b) To obtain the histogram for the Math SAT scores for males using a class interval size of 100, we merely press: WINDOW and change Xscl to 100 and Ymax to 150000. We then press GRAPH . To determine the number of scores in the range $500 \le x < 600$, we press: TRACE and move the cursor to this interval. We obtain the display on the right. From this display, we see that there were 129,508 scores between 500 and 600.

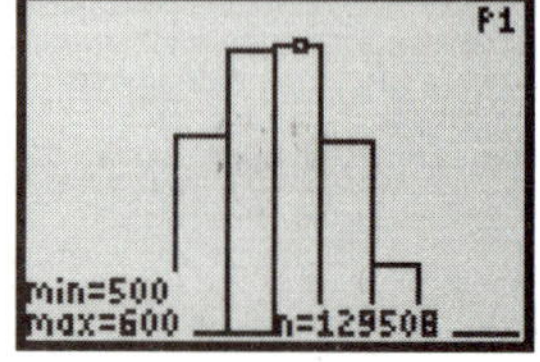

WORKED EXAMPLE 2

a) Create a list, L_4, which gives the *cumulative* frequencies of the SAT scores for males. That is, $L_4(n)$ is the total frequency of all scores in the first n class intervals where the nth class interval is defined to be the interval: $200 + 50(n - 1) \leq x < 200 + 50n$: $n > 0$.

b) Construct an xyLine showing the percentage of all males whose scores were in the first n class intervals. Trace along your xyLine to find the percentage of males whose Math SAT scores were less than 600.

Solution

a) We redefine L_1 to replace the midpoints of the class intervals with their upper bounds:
   $L_1 = \{250, 300, 350, 400, 450, 500, 550, 600, 650, 700, 750, 800\}$

The corresponding frequencies of scores (for males) in these intervals are given by L_2. The *cumulative* frequencies are a running total, L_4, of the numbers in list L_2. To compute the elements of L_4, we start with the first term of L_2, 4117. That is, $L_4(1) = 4117$. To get $L_4(2)$, we add $L_2(2)$, to $L_4(1)$ so that $L_4(2) = L_4(1) + L_2(2)$ or $4117 + 19581$. In general, we can generate L_4 using the equation: $L_4(n) = L_4(n-1) + L_2(n)$.

To define L_4 recursively, we access **sequence** mode[1] by pressing: **MODE**
Then we select **seq** from the fourth row of that display and press: **ENTER** **Y =**
The screen displays the sequence variables U_n and V_n.
We represent the list (sequence), L_4 by U_n and define U_n recursively by $U_n = U_{n-1} + L_2(n)$.
To assign the 12 terms of U_n to the list L_4, we press:

2nd [LIST] **5** **2nd** [Y-VARS] **4** **ENTER** **(** **X,T,θ** **)** **,**

X,T,θ **,** **1** **,** **1** **2** **,** **1** **)** **STO►** **2nd** **[L₄]** **ENTER**

They keying sequence above yields the display shown here. ⸺

Upon pressing: **STAT** **ENTER** **►** **►** **►** we obtain the table showing L_4.

b) To create a list, L_5 such that $L_5(n)$ is the *percentage* of marks in the first n class intervals, we return to the home screen and press: **1** **0** **0** **x** **2nd** **[L₄]**
÷ **2nd** [LIST] **►** **5** **2nd** **[L₂]** **STO►** **2nd** **[L₅]** **ENTER**

To plot the xyLine with L_1 as the **Xlist** and L_5 as the **Ylist**, press:

2nd [STAT PLOT] **4** **ENTER** This turns off all the statistical plots. Then press: **2nd** [STAT PLOT]

We select Plot 3 and then we choose the second icon in the "Type" row (to plot the xyLine) and select all other settings as shown in the display below left. To set the window variables so that the entire xyLine fits in the window, and to plot the xyLine, press: **ZOOM** **9** .

This yields the graph shown in the display on the right. When we press: **TRACE** and move the cursor along the curve to $x = 600$, the display shows **Y** $= 75.60\ldots$ indicating that about 75.6% of all the SAT math scores by males were less than 600. Such a cumulative percentage curve is called an *ogive curve*.

Frequency   **Cumulative Frequency**

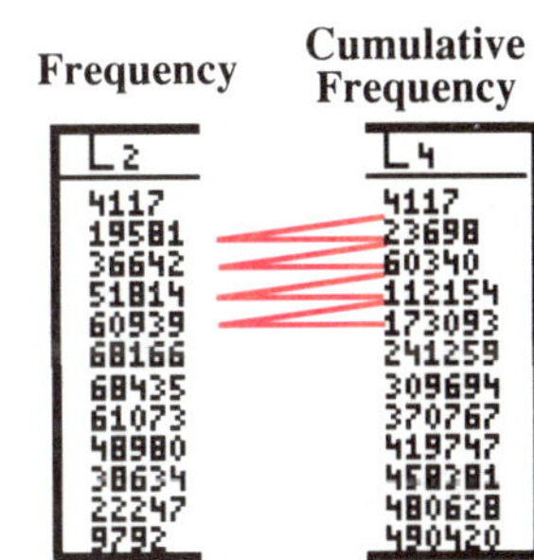

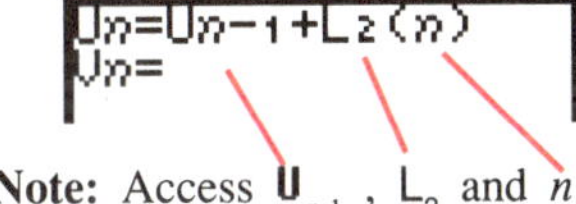

Note: Access U_{n-1}, L_2 and n through the 2ⁿᵈ function keys.

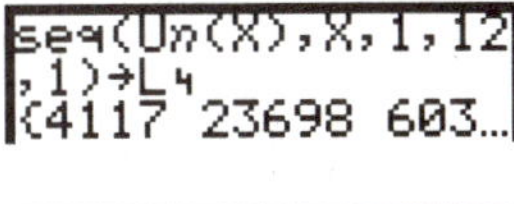

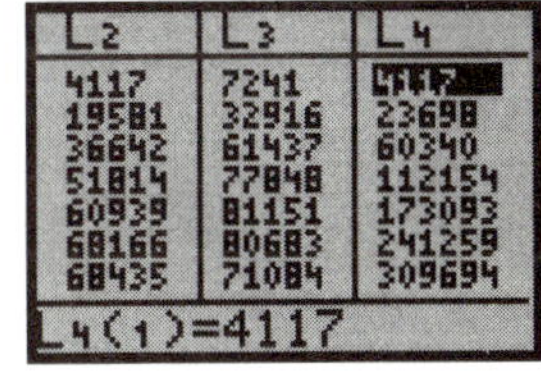

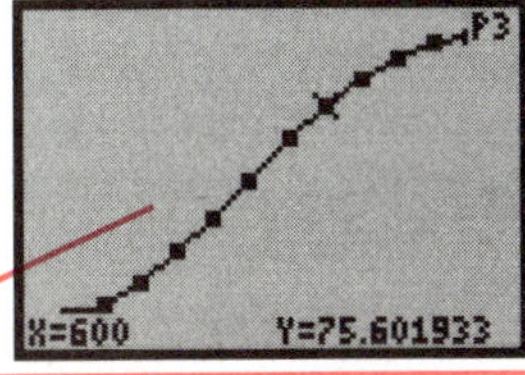

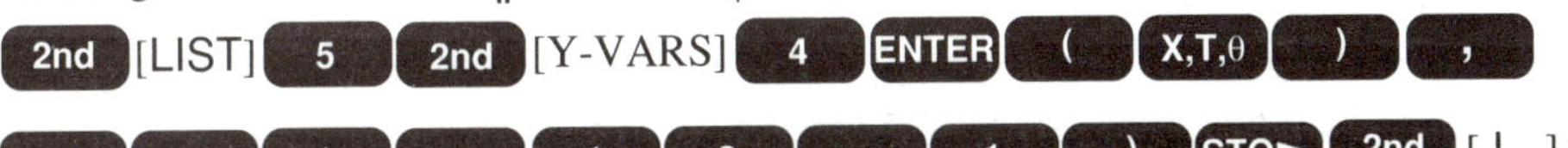

---

[1] For a detailed treatment of sequences, see pages 16ff of our publication, *Programming and Programs for the TI-82 Graphics Calculator*.

Exercises

1. Use the histogram on page 9 showing the profile of math SAT scores for males to answer the following questions.
- a) How many males scored below 250 on the SAT?
- b) How many males scored below 400?
- c) What percent of male respondents scored below 400?
- d) What percent of males scored 700 or above?
- e) Can you determine from the histogram or table what percent of the male respondents scored above 700? Explain your answer.

2. a) Follow the procedures in worked example 1 to re-create the table shown in that example. Check that the numbers in columns L_1, L_2 and L_3 of your table agree with the numbers in the given table.

b) Define L_4 so that $L_4 = L_2 + L_3$. Use your table from part a), and the procedures outlined in worked example 1, to create a histogram showing the total frequencies of the SAT scores of males and females combined. Use the same intervals as in the table. (These are called the *class intervals*.)

c) Change your plot from a histogram to an xyLine by selecting that option from the appropriate plot on the STAT PLOTS screen. This graph is called a *frequency polygon*.

d) Change the histogram so the class interval size is 100; i.e. $200 \leq x < 300$, $300 \leq x < 400, \ldots\ 700 \leq x < 800$. What proportion of the students had SAT scores in the 500s?

e) What proportion of the students scored less than 600?

3. The table below shows the number of golfers on the PGA Tour in 1993 whose total winnings were in each $100,000 dollar range between $200,000 and $1,500,000. (Nick Price was at the top with winnings of $1,478,557.)

L_1 is the list of lower limits (in dollars) of the class intervals. L_2 is the number of golfers in the interval with lower limit given in L_1. For example, the first row of the table shows that there were 31 golfers whose 1993 money winnings were in the range:
$$\$200,000 \leq x < \$300,000.$$

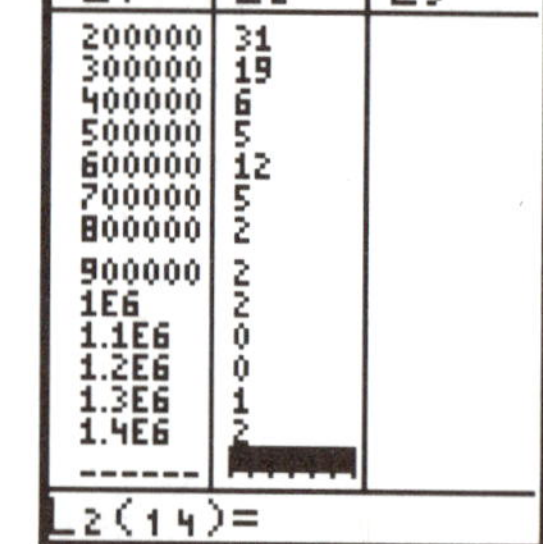

L1	L2	L3
200000	31	
300000	19	
400000	6	
500000	5	
600000	12	
700000	5	
800000	2	
900000	2	
1E6	2	
1.1E6	0	
1.2E6	0	
1.3E6	1	
1.4E6	2	

L2(14)=

a) Create a table in which L_1 gives the midpoints of the class intervals. Display these data in a histogram. Construct the corresponding frequency polygon. (See exercise 2c.)

b) What percentage of the over $200,000 money winners on the PGA Tour won more than half a million dollars?

c) Change the histogram so the class interval size is 200 000. What percentage of the golfers in the top 130 money winners had 1993 winnings in the interval $500,000 \leq x < \$700,000$?

Investigations

4. Define list L_1 in your STAT table as in worked example 2. Define L_2, L_3 and L_4 as in exercise 2. Then define L_5 so that it gives the cumulative frequencies of the SAT scores for all students. (Follow the procedures in worked example 2.)

a) Define L_6 so that it gives the cumulative percentages of students scoring below any upper bound. Construct the cumulative percentage curve (ogive curve) for the SAT math scores of all students.

b) Use your ogive curve to determine the percent of all students whose Math SAT scores were less than 600. Compare this answer with your answer to part e) of exercise 2.

c) In what interval is the ogive curve steepest? Explain why.

5. Construct the ogive curve for the money winnings of the top golfers as given in the table of exercise 3. Use a class interval size of 100 000. Remember to define L_1 as the list of upper bounds in the class intervals.

a) Trace along the ogive curve to determine what percent of golfers in the "$200,000 plus" category won more than half a million dollars in 1993. Compare your answer to your answer to part b) of exercise 3.

b) Describe the shape of the ogive curve. Explain why it flattens out for large values of x.

Note: When creating the cumulative frequencies, L_3, remember to select **sequence** mode and define U_n as shown in worked example 2. Also set U_n**Start=0** and set n to run from 1 to 13 (number of class intervals). Define L_3 using the assignment **seq(U$_n$(H), H, 1, 13, 1)→L$_3$**.

DRAWING INFERENCES

A prestigious university selects only those students whose Math SAT scores are in the top 10% of all those who write this test. Estimate what score a student would have needed in 1990 to satisfy this university's entrance requirement.

In view of the differential between the performance of males and females on this test, could the selection criterion described above be considered sexist (i.e. biased against females)? Explain your answer.

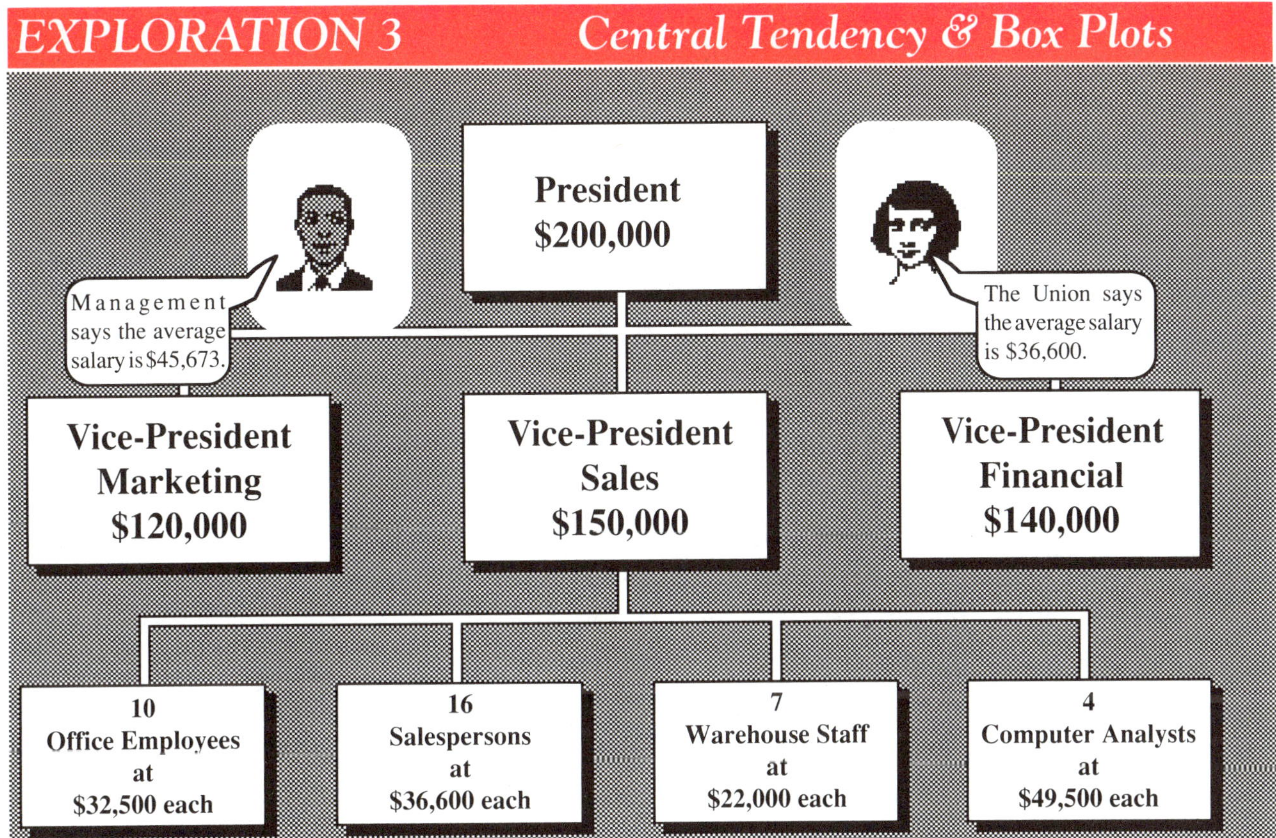

The organization chart for High Tech Widget Inc. is presented above. It shows the number of employees in each position and the corresponding annual salary. During salary negotiations, management asserts that the average salary of the workers is $45,673. The Union asserts that the average salary is only $36,600. Who is telling the truth?

Upon closer investigation, it was discovered that the two sides were using different definitions of average. Management used the *mean*, which is the total money which the company pays in salary, divided by the number of employees. This is the amount of money which each employee would earn if all employees were paid equally. In general, the *mean* (denoted $\bar{x}$) of a set of n numbers, $\{x_1, x_2, x_3, \dots x_n\}$ is:

$$\bar{x} = \frac{x_1 + x_2 + x_3 + \dots + x_n}{n}$$

The Union was using the *median* salary as their measure of the average. If the salaries of the 41 employees at High Tech Widget Inc. are arranged in descending order from the president to the lowest paid worker, the *median* salary is the 21st salary in that list. There are the same number of workers earning as much or more than this salary as there are earning as much or less than this salary. In general, the median of a set of n numbers, $\{x_1, x_2, x_3, \dots x_n\}$ is the middle number when they are arranged in ascending order. If n is even, there are two "middle numbers" and the median is taken to be half their sum.

WORKED EXAMPLE

a) Display the salaries of High Tech Widget Inc. in a table showing the number of people earning each salary amount.

b) Find the mean salary for High Tech Widget Inc.

c) Find the median salary. Define the lower and upper quartiles and calculate these quartiles for the list of salaries. Display the lower and upper quartiles and the median salary on a box plot..

Solution

a) The **first step** in calculating any statistics is to enter the data in a table.
To do this, we press: **STAT** **ENTER** .

We then enter the salaries in L_1 and their corresponding frequencies in L_2 to obtain the table shown on the right.

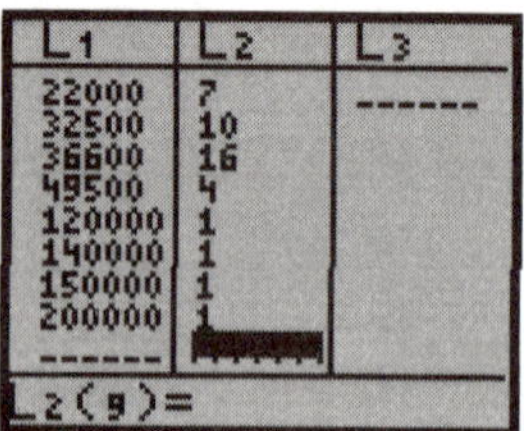

b) The **second step** in calculating a statistic is to access the STAT CALC menu by pressing:

STAT **▶** . We then select the **SetUp** menu by pressing: **3**

We obtain the display shown. Since there is only one variable (salary), we select one variable statistics, (**1-Var Stats**) and highlight L_1 for the **Xlist** and L_2 for the frequency.

We then perform the **1-Var Stats** calculations by pressing: **STAT** **▶** **ENTER** **ENTER**.

The display is shown below left. We see that the mean, $\bar{x}$, is equal to $\bar{x} = 45,673.17$. That is, the mean salary is about $46,673 as stated by the management.

c) The arrow on the display showns that we can scroll downward to obtain the display on the right. This display shows that there are 41 salaries ($n = 41$) and that the salaries range from $22,000 (**minX=22000**) to $200,000 (**maxX=200000**).
The median salary is $36,600 (**Med=36600**) as stated by the Union.

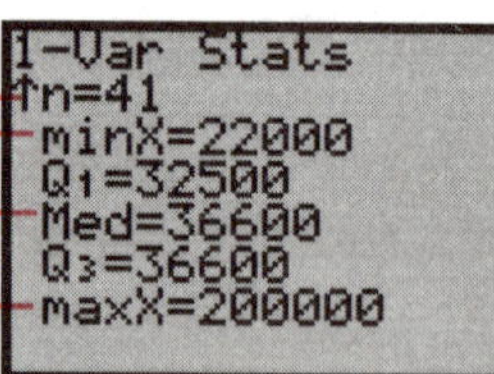

The symbols, Q_1 and Q_3 represent the *first quartile* and *third quartile* respectively.
The first quartile is the salary below which one quarter of the salaries lie. That is, if the 41 salaries were arranged in ascending order, the mean of the 10th and 11th salaries would be the first quartile. Similarly, the mean of the 30th and 31st salaries would be the third quartile. In general, one quarter of all the values in a set of data are at or below the value of the first quartile, one half of the values are at or below the median and three-quarters are at or below the third quartile. The display shows that the first quartile is $32,500 and the third quartile is $36,600. (Why are the median and the third quartile equal?)

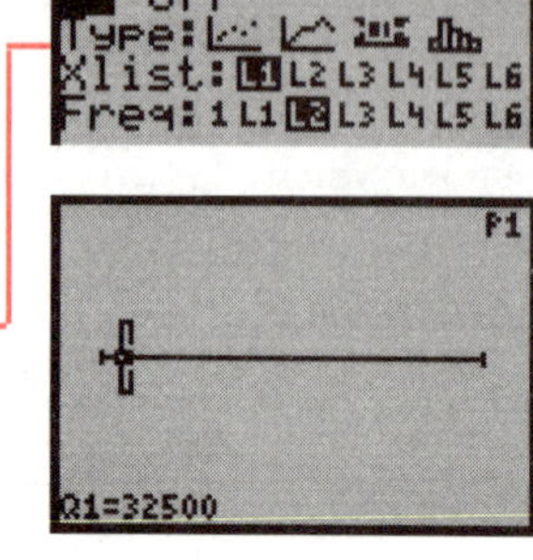

To create a box plot, press: **2nd** [STAT PLOT] **4** **ENTER** . This turns off all the data plots.
To access plot1 press: **2nd** [STAT PLOT] **ENTER** Then select the options shown in the display.

Upon pressing **ZOOM** **9** we obtain the box graph shown on the right. By tracing along this graph we can verify the values of the quartiles and the extremes shown above.

Exercises

1. Why are measures of central tendency such as the mean and the median used?

2. Suppose a single position paying $180,000 were created at High Tech Widget Inc. Which average salary, the mean or the median would be increased more by the inclusion of this new salary?

3. Use your calculator to find the mean, median, first quartile and third quartile of each of the following sets of numbers. (One way to clear a previous list is to place the cursor on the list name in the STAT table and press: `CLEAR` `ENTER` .)

a) {15, 16, 29, 36, 45, 16, 18, 29, 16, 29, 32, 36}
b) {-17, 61, 42, 156, -88, -61, 42, 42, 61, -17, -17}

4. *How many dimples on a golf ball?* Most people do not realize (and some do not care) that a golf ball without dimples would not travel very far. The dimples give it aerodynamic properties which increase the distance it travels. Engineers experiment to determine the size and number of dimples which yield optimum distance. A recent issue of a popular golf magazine published the number of dimples on 32 of the most popular brands of golf ball. (shown in the table below)

Number of Dimples on each Golf Ball							
440	432	392	432	318	442	432	432
432	360	360	360	432	492	440	392
392	492	332	422	422	422	332	440
392	332	432	432	392	332	392	360

Find the mean and median number of dimples on the 32 golf balls given in the table above…
a) by entering each of the 32 numbers into list L_1 in the STAT editor and using the 1-Var Stats as in the worked example.
b) by making a tally like this and using the procedure in the worked example.

318	332	360	392	422	432	440	442	492
/	////				etc. …			

Check that n = 32 on the **1-Var Stats** display to ensure that you have entered all the data points.
c) Construct a box plot to show the quartiles and extremes.

5. During the 1970's the mean income in a southern state increased by 115% for non-whites and 72% for whites. About 80% of the people in this work force were white. What was the mean increase in income for both groups combined?

6. Determine the mean and median of each set of numbers.
a) {1, 2, 3, 4, …99} (the natural numbers from 1 to 99)
b) {1, 2, 3, 4, …100} (the natural numbers from 1 to 100)
c) {-8, -7, -6, -5, -4, -3, -2, -1, 0, 1, 2, 3, 4, 5}

Investigations

7. The table of SAT scores on page 9 gives the number of students who achieved scores in various intervals. Since the *exact* scores of individual students are not known, we cannot compute the median score from this information. However, we can approximate the median score if we assume each student received the middle score in each interval. That is, we assume the 4117 male students and 7241 female students who achieved scores in the interval $200 \leq x \leq 250$ all achieved scores of 225.
Making this assumption, construct box plots showing the quartile and extreme scores for the males and for the females on the SAT. Compare the shapes of the two box plots.

8. In a study of the longevity of a particular species of cat, biologists recorded the lifespans of 30 cats. Their results are presented in the following table.

Lifespans of Cats (in years)									
12.9	13.2	14.1	13.9	12.8	13.1	13.1	13.2	13.6	13.0
13.4	13.6	12.9	13.3	11.8	12.8	14.6	12.8	10.4	14.8
11.5	13.5	13.6	12.9	9.6	14.5	13.5	13.8	14.4	13.3

In order to enter these data into a list, it is useful to display them in a stem-and-leaf diagram in which the stem is the whole number and the leaf is the decimal digit.

Stem	Leaf	
9	6	This stands for 9.6
10	4	
11	5 8	This stands for 11.8
12	8 8 8 9 9 9	
13	0 1 1 2 2 3 3 4 5 5 6 6 6 8 9	
14	1 4 5 6 8	

Use the information in the stem-and-leaf plot to find the mean and median lifespans in years. Create a box plot and trace to find the first and third quartiles.

CHALLENGE

A doctor lists all the diseases which afflict one or more of his patients. Opposite each disease he records its frequency among his patients. That is, he records the number of patients which have each of these diseases.

The set of frequencies for all the diseases on his list has an arithmetic mean of 20. If the flu, which has a frequency of 92 is removed from the list, the arithmetic mean of the remaining frequencies is 18.

How many diseases remain on the doctor's list ? What is the largest possible frequency for any disease which remains on the list?

Do Great Actresses Mature Earlier Than Great Actors?

Questions like the one above are difficult to answer because the answer usually depends upon the individuals who are compared. Such questions are often rephrased as follows, "Does the average actress reach maturity faster than the average actor?". Again the question is meaningless because it assumes the existence of a fictional entity called an "average actress". The following question is better defined and more measurable, though it does not address the original question: "How does the average age of the oscar winners for best actor compare with the average age of the oscar winners for best actress?

The tables on pages 16 and 17 show the ages of the winners of the "best actress" and "best actor" awards in the 50 years from 1942 through 1991.

"BEST ACTRESSES" DURING A HALF CENTURY— 1942-1991

Year	Actress	Age	Movie
1942	Greer Garson	34	Mrs. Minever
1943	Jennifer Jones	24	The Song of Bernadette
1944	Ingrid Bergman	29	Gaslight
1945	Joan Crawford	41	Mildred Pierce
1946	Olivia de Havilland	30	To Each His Own
1947	Loretta Young	34	The Farmer's Daughter
1948	Jane Wyman	34	Johnny Belinda
1949	Olivia de Havilland	33	The Heiress
1950	Judy Holliday	28	Born Yesterday
1951	Vivien Leigh	38	Streetcar named Desire
1952	Shirley Booth	45	Come Back Little Sheba
1953	Audrey Hepburn	24	Roman Holiday
1954	Grace Kelly	26	The Country Girl
1955	Anna Magnani	48	The Rose Tattoo
1956	Ingrid Bergman	41	Anastasia
1957	Joanne Woodward	27	The Three Faces of Eve
1958	Susan Hayward	40	I Want to Live
1959	Simone Signoret	38	Room at the Top
1960	Elizabeth Taylor	28	Butterfield 8
1961	Sophia Loren	27	Two Women
1962	Anne Bancroft	31	The Miracle Worker
1963	Patricia Neal	37	Hud
1964	Julie Andrews	30	Mary Poppins
1965	Julie Christie	24	Darling
1966	Elizabeth Taylor	34	Who's Afraid of Virginia Woolf?

Year	Actress	Age	Movie
1967	Katharine Hepburn	60	Guess Who's Coming to Dinner?
1968 (double award)	Katharine Hepburn	61	The Lion in Winter
1968 (double award)	Barbara Streisand	26	Funny Girl
1969	Maggie Smith	35	The Prime of Miss Jean Brodie
1970	Glenda Jackson	34	Women in Love
1971	Jane Fonda	34	Klute
1972	Liza Minelli	26	Cabaret
1973	Glenda Jackson	37	A Touch of Class
1974	Ellen Burstyn	42	Alice Doesn't Live Here Anymore
1975	Louise Fletcher	41	One Flew Over the Cuckoo's Nest
1976	Faye Dunaway	35	Network
1977	Diane Keaton	31	Annie Hall
1978	Jane Fonda	41	Coming Home
1979	Sally Field	33	Norma Rae
1980	Sissy Spacek	30	Coal Miner's Daughter
1981	Katharine Hepburn	74	On Golden Pond
1982	Meryl Streep	33	Sophie's Choice
1983	Shirley MacLaine	49	Terms of Endearment
1984	Sally Field	38	Places in the Heart
1985	Geraldine Page	61	Trip to Bountiful
1986	Marlee Matlin	21	Children of a Lesser God
1987	Cher	41	Moonstruck
1988	Jodie Foster	26	The Accused
1989	Jessica Tandy	80	Driving Miss Daisy
1990	Kathy Bates	42	Misery
1991	Jodie Foster	29	The Silence of the Lambs

[1]This exploration is based on an article by Richard Brown and Gretchen Davis, published in *Mathematics Teacher*, February, 1990 issue.

"BEST ACTORS" DURING A HALF CENTURY— 1942-1991

Year	Actor	Age	Movie
1942	James Cagney	43	Yankee Doodle Dandy
1943	Paul Lukas	48	On the Rhine
1944	Bing Crosby	43	Going My Way
1945	Ray Milland	40	The Lost Weekend
1946	Fredric March	49	Best Years of our Lives
1947	Ronald Colman	56	A Double Life
1948	Laurence Olivier	41	Hamlet
1949	Broderick Crawford	38	All the King's Men
1950	Jose Ferrer	38	Cyrano de Bergerac
1951	Humphrey Bogart	52	The African Queen
1952	Gary Cooper	51	High Noon
1953	William Holden	35	Stalag 17
1954	Marlon Brando	30	On the Waterfront
1955	Ernest Borgnine	38	Marty
1956	Yul Brynner	41	The King and I
1957	Alex Guinness	43	Bridge on the River Kwai
1958	David Niven	49	Separate Tables
1959	Charleton Heston	35	Ben Hur
1960	Burt Lancaster	47	Elmer Gantry
1961	Maximillian Schell	31	Judgment at Nuremburg
1962	Gregory Peck	46	To Kill a Mockingbird
1963	Sidney Poitier	39	Lillies of the Field
1964	Rex Harrison	56	My Fair Lady
1965	Lee Marvin	41	Cat Ballou
1966	Paul Scofield	44	A Man for All Seasons

Year	Actor	Age	Movie
1967	Rod Steiger	42	In the Heat of the Night
1968	Cliff Robertson	43	Charly
1969	John Wayne	62	True Grit
1970	George C. Scott	43	Patton
1971	Gene Hackman	40	The French Connection
1972	Marlon Brando	48	The Godfather
1973	Jack Lemmon	48	Save the Tiger
1974	Art Carney	56	Harry and Tonto
1975	Jack Nicholson	38	One Flew Over the Cuckoo's Nest
1976	Peter Finch	60	Network
1977	Richard Dreyfuss	32	The Goodbye Girl
1978	Jon Voight	40	Coming Home
1979	Dustin Hoffman	42	Kramer vs. Kramer
1980	Robert de Niro	37	Raging Bull
1981	Henry Fonda	76	On Golden Pond
1982	Ben Kingsley	39	Ghandhi
1983	Robert Duvall	55	Tender Mercies
1984	F. Murray Abraham	45	Amadeus
1985	William Hurt	35	Kiss of the Spider Woman
1986	Paul Newman	61	Color of Money
1987	Michael Douglas	33	Wall Street
1988	Dustin Hoffman	51	Rainman
1989	Daniel Day Lewis	32	My Left Foot
1990	Jeremy Irons	42	Reversal of Fortune
1991	Anthony Hopkins	54	The Silence of the Lambs

WORKED EXAMPLE

Display the ages of the "best actress" and the "best actor" winners in a back-to-back stem-and-leaf plot. Create and trace the box plots for these ages to find the age quartiles for "best" actors and actresses.

Solution

Back-to-Back Stem-and-Leaf Plot

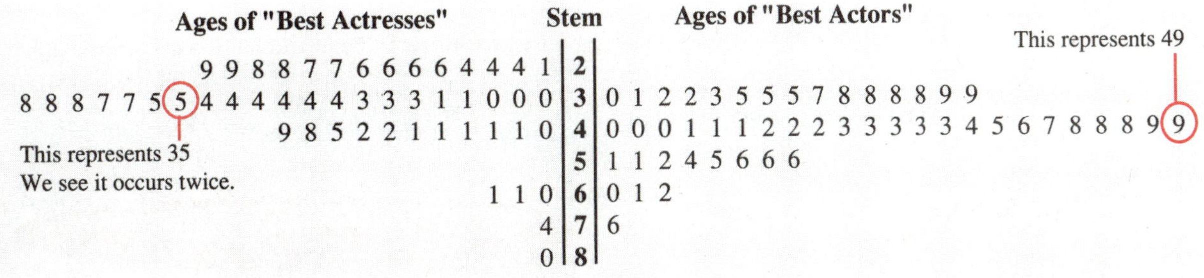

To create a box plot, for the ages of the actresses and actors, follow these steps:

❶ Enter the ages of "best actresses" and "best actors" into lists, L_1 and L_3 respectively and the corresponding frequencies into L_2 and L_4.

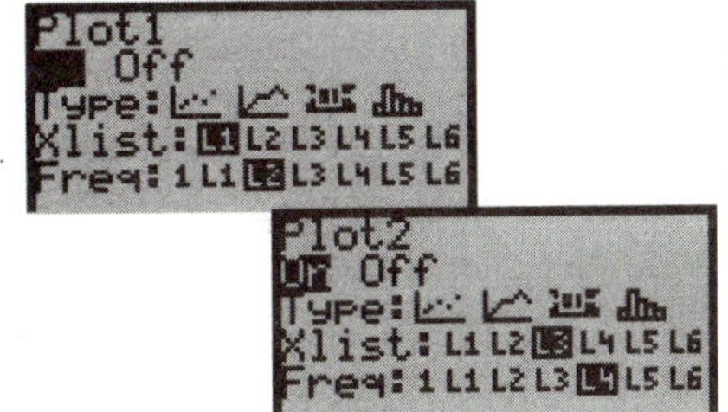

L_1	L_2	L_3	L_4
21	1	30	1
24	3	31	1
26	4	32	2
27	2	33	1

❷ Press: **2nd** [STAT PLOT]. Then define Plot 1 and Plot 2 as shown below.

❸ Press: **ZOOM** **9** to obtain the box plots shown below.

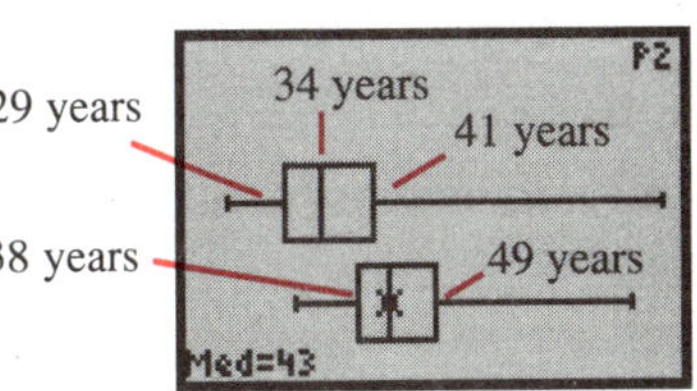

We trace along the box plots to find the quartiles shown in the display.

1. Follow the procedures in the worked example to create the box plots showing the age distributions of "best actress" and "best actor" award winners.

a) Compare the median age of the "best actor" award winners with the median age of "best actress" award winners.

b) Why might you expect the "best actress" award winners to have a younger median age than the "best actor" award winners?

c) Trace along your box plots to find the lower and upper extreme (minimum and maximum) ages of the award winners.

(Note: To move from one box plot to the other, press: )

d) Compare the lower extreme age of the "best actresses" with the lower extreme age of the "best actors". Why might you expect the lower extreme age of the "best actress" winners to be less than the lower extreme age of the "best actor" winners?

e) Compare the upper extreme age of the "best actresses" with the upper extreme age of the "best actors". Why might you expect the upper extreme age of the "best actress" winners to be close to the upper extreme age of the "best actor" winners?

f) The difference between the upper and lower extremes is called the *range*. Calculate and compare the ranges for the "best actors" and for the "best actresses".

g) The difference between the first quartile, Q_1 and the third quartile, Q_3 is called the *interquartile range*. Calculate and compare the interquartile ranges for the "best actors" and for the "best actresses".

h) Which difference do you think will be greater; the differences in the median ages of the "best actor" and "best actresses", or the difference in their means? Explain. Access the **1-Var Stats** using the keying sequence described in the worked example on page 14 to find the mean ages of the "best actress" and "best actor" award winners.

2. Box plots like the one in the worked example are sometimes called *box-and-whisker* plots. The box is the rectangular region bounded by Q_1 and Q_3 and the whiskers are the horizontal lines extending from the box to the extremes.
a) What percent of the data are represented by values inside the box?
b) What percent of the data are represented by values between and including the ends of the whiskers?
c) When does the median occur at the center of the box?
d) Describe a data set which has a narrow box and long whiskers?

3. Is there a trend toward selecting older women for the "best actress" award?
a) Construct a box plot for the ages of the "best actresses" spanning the quarter century from 1942 to 1966.

b) On the same display, construct a box plot for the ages of the "best actresses" for the quarter century from 1967 to 1991.

c) Compare the box plots in parts a) and b). Does there appear to be a trend toward selecting older women? Explain.

d) Construct box plots for the ages of the "best actors" over the two quarter centuries 1942-66 and 1967-91. Is there a trend toward selecting older men for the "best actor award?

4. a) The *mode* in a set of data is the value which occurs most often. What is the modal age for the "best actress" award winners over the period 1942-91?

 b) What is the modal age for the "best actor" award winners over the period 1942-91?

5. We noted in exercise 1g) that the length of the box in the box plot of any set of data is called the *interquartile range*. Any value in a set of data whose distance from the box exceeds 1.5 times the interquartile range is called an *outlier*.

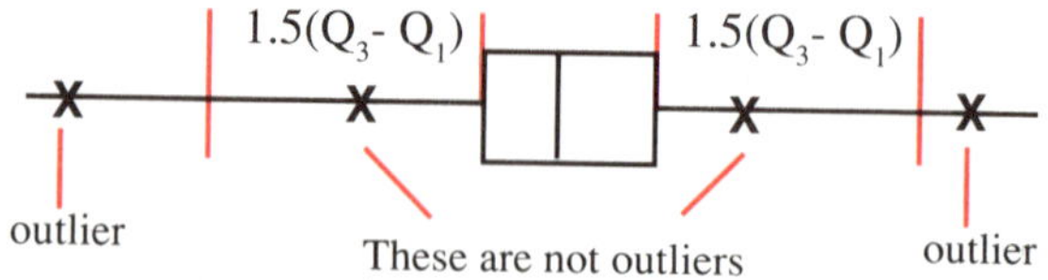

a) List any outliers in the set of ages of "best actress" winners.
b) List any outliers in the set of ages of "best actor" winners.
c) Explain the significance of an outlier in the set of ages of award winners.

CHALLENGE

In a set of data, $\{x_1, x_2, x_3, \ldots x_n\}$, the deviation of x_i from the average is given by $|x_i - x_{average}|$. Which value for $x_{average}$, the mean or the median, minimizes the sum of the deviations for all x_i?

To calculate the sum of the deviations for the ages of the actresses using $x_{average} = \bar{x}$, define L_5 by abs$(L_1 - \bar{x}) \to L_5$ and then $L_5 \times L_2 \to L_6$. Then evaluate **sum** L_6. Repeat the calculation replacing $\bar{x}$ with Med to evaluate the sum of the deviations $|x_i - \text{Med}|$.

In Explorations 3 and 4, we studied two different measures of central tendency, the mean and the median. If we imagine a set of numbers $x_1 < x_2 < x_3 < \ldots < x_6$ distributed along a number line, the *mean* would correspond to the center of gravity or balance point of the set.

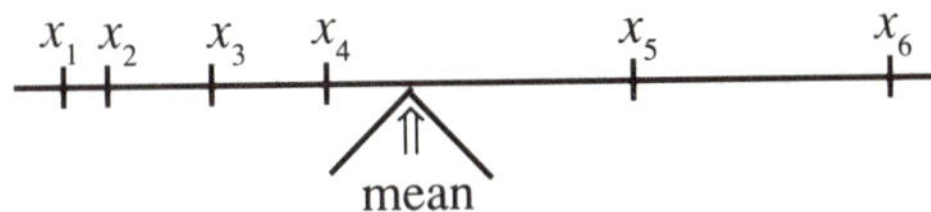

That is, the sum of the differences $x_i - \overline{x}$ is 0. Adding any two data points to our set does not change the mean if the two new values are equidistant from the mean and on opposite sides.

Similarly, the *median* in this set corresponds to the middle value (or the arithmetic mean of the two middle values). Adding two new data points to the set does not change the median provided the two new points are on opposite sides of the median. In short, measures of central tendency tell us the "mid point" of a set of data but they do not tell us how closely the values cluster around this mid point. An old adage says:

> *If your head is in the furnace and you're standing on*
> *a block of ice, your average body temperature is normal.*

In the CHALLENGE investigation of Exploration 4, you may have touched upon the following interesting properties of the mean and the median.

> Consider all the distances $|x_i - x_0|$ between the points in the data set and a particular point, x_0. The sum of these distances is a minimum when x_0 is the median and the sum of the squares of these distances is a minimum when x_0 is the mean

The quartiles and the corresponding box plots studied in the previous explorations, provide some measure of the distribution of the data around the median. To provide another measure of the extent to which the data "cluster" around the mean, statisticians have developed measures of *dispersion*. The most widely used measure of dispersion is the *standard deviation* (usually denoted by the Greek letter, sigma, σ). The standard deviation of the set of data $\{x_1, x_2, x_3, \ldots x_n\}$ is defined by

$$\sigma = \sqrt{\frac{(x_1 - \overline{x})^2 + (x_2 - \overline{x})^2 + \ldots\ldots\ldots + (x_n - \overline{x})^2}{n}}$$

Although this formula appears complicated, it is composed as follows: We square all the *deviations from the mean*, $|x_i - \overline{x}|$, (i. e. we square the "distance" of each x_i from the mean) and compute the average of these. Then take the square root to get σ. The squaring of the deviations from the mean makes σ more sensitive to extreme values in the data. If the gentleman in the cartoon had insisted on batteries whose lifetimes have a small standard deviation, his watch would probably have run for close to 5.7 years.

WORKED EXAMPLE

The lifetimes (in years) of 30 Brand A and 30 Brand B batteries are given in the tables below. Which of these brands of battery is more reliable?

Measured Lifetimes of 30 Brand A Batteries				
5.1	7.3	6.9	4.7	4.6
6.2	6.4	5.5	4.9	6.9
6.0	4.8	4.1	5.3	8.1
6.3	7.5	5.0	5.7	9.3
3.3	3.1	4.3	5.9	6.6
5.8	5.0	6.1	4.6	5.7

Measured Lifetimes of 30 Brand B Batteries				
5.4	6.3	5.0	5.9	5.6
4.7	6.0	3.3	6.6	6.0
5.0	6.5	5.8	5.4	4.9
5.7	6.8	5.6	4.9	6.0
4.9	5.7	6.2	7.5	5.8
6.8	5.9	5.3	5.6	5.9

Solution

To compare the distributions of the lifetimes of the two brands we may proceed as in Exploration 4 and create boxplots for the lifetimes of each brand. We organize our data in a stem-and-leaf plot.

Back-to-Back Stem-and-Leaf Plot

Brand A	Stem	Brand B
3 1	3	3
9 8 7 6 6 3 1	4	7 9 9 9
9 8 7 7 5 3 1 0 0	5	0 0 3 4 4 6 6 6 7 7 8 8 9 9 9
9 9 6 4 3 2 1 0	6	0 0 0 2 3 5 6 8 8
5 3	7	5
1	8	
3	9	

To create box plots, for the lifetimes of brands A and B, we follow these steps:

❶ Enter the lifetimes of brands A and B into lists, L_1 and L_3 respectively and the corresponding frequencies into L_2 and L_4.

❷ Press: **2nd** [STAT PLOT]. Then define Plot 1 and Plot 2 as shown below.

❸ Press: **ZOOM** **9** to obtain the box plots shown below.

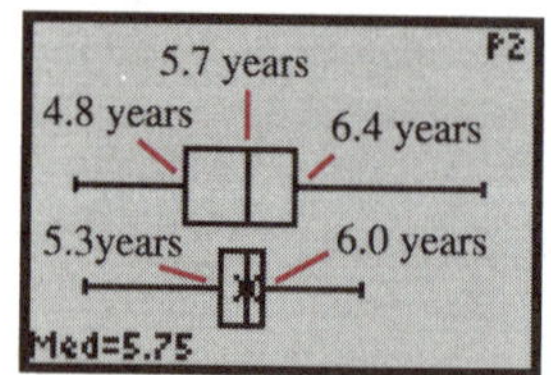

We trace along the box plots to find the quartiles shown in the display.

We observe from the box plots that both batteries have about the same median. However the interquartile range (length of the box) for Brand A batteries (upper box plot in the display) is much greater than the interquartile range for the Brand B batteries. Therefore the Brand B batteries are more reliable. We can also compare reliabilities by computing $\bar{x}$ and σ. To do this, we press: **STAT** ▶ **3** to access the STAT SET UP menu. Under **1-Var Stats**, we select L_1 for the Xlist and L_2 for the frequency. To calculate the **1-Var Stats**, we press: **STAT** ▶ **ENTER** **ENTER**, and obtain the display for Brand A. Similarly, we obtain the display for Brand B. Both brands of battery have a mean life of 5.7 years; however, the standard deviation, σ_x is about 1.33 years for Brand A, and only about 0.78 for Brand B, so Brand B is more reliable.

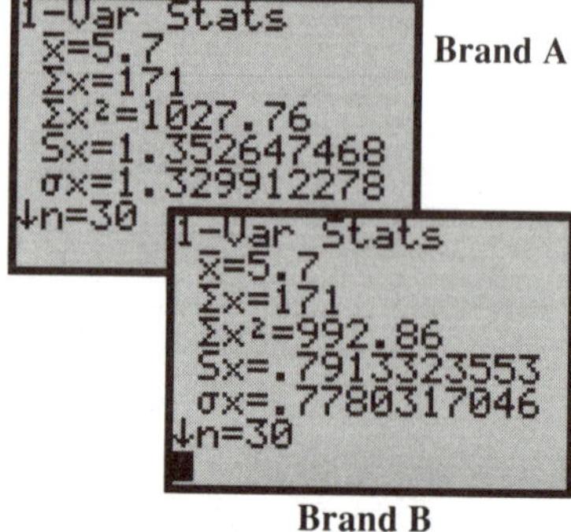

1. Why did statisticians develop measures of dispersion?

2. The simplest measure of dispersion is called the *range*. The *range* of a set of data is defined to be the difference between the largest and smallest values. In the worked example:

a) What is the range of the lifetimes of the Brand A batteries?
b) What is the range of the lifetimes of the Brand B batteries?
c) Why is the standard deviation used more often than the range as a measure of dispersion?

3. Another measure of dispersion, called the *mean deviation* is designed to take into account the deviation of each number in a set of data from the mean. The *mean deviation* is defined to be the mean of all the "distances" between the x_i and $\bar{x}$. That is, the mean deviation of a set of data, $\{x_1, x_2, x_3, \ldots x_n\}$

is:
$$\frac{|x_1 - \bar{x}| + |x_2 - \bar{x}| + |x_3 - \bar{x}| + \ldots |x_n - \bar{x}|}{n}$$

(Vertical bars denote absolute value.)
a) Determine the mean deviation in the lifetimes of Brand A batteries. (Do the subtractions in your head and enter the values $|x_i - \bar{x}|$ into the calculator to form a running total.)

b) Compute the mean deviation for the Brand B batteries.

c) Compare your answers in parts (a) and (b) and determine which brand of battery is more reliable according to this measure of dispersion.

4. Calculate the mean deviation and the standard deviation for
a) $S = \{1, 2, 3, 4, 5, 6, 7, 8, 9, 10, 11\}$
b) $T = \{-5, -4, -3, -2, -1, 0, 1, 2, 3, 4, 5\}$
What is the effect on the standard deviation of a set of data when a fixed constant is added to all the values?

5. Compare the mean deviation (see exercise 3) and the standard deviation as measures of dispersion. Which measure is more sensitive to data which has some large deviations from the mean?

6. The reputation of a steel processing plant rests on its ability to manufacture products within fine tolerance limits. Samples of 30 ball bearings which are advertised as 8 mm in diameter by Acme Bearings were measured and compared with a sample of 30 produced by Delta Bearings. Calculate $\bar{x}$ and σ for the samples below, to determine which company produces bearings with the smaller deviations from 8 mm.

Diameters of 30 Ball Bearings from Acme					Diameters of 30 Ball Bearings from Delta				
7.93	7.87	7.84	8.09	8.14	8.01	8.04	8.09	7.98	8.05
8.28	7.88	8.01	7.92	7.98	8.10	8.05	8.09	7.99	7.99
8.01	8.05	8.01	8.07	7.94	7.97	7.99	8.00	8.03	8.03
8.03	8.14	8.09	7.98	8.01	7.94	7.97	8.03	7.93	8.10
7.93	7.98	8.09	7.95	8.04	8.00	8.03	8.06	8.08	8.11
8.13	7.90	7.89	8.18	8.13	8.14	7.96	8.05	8.08	8.00

7. One of the histograms below displays the lifetimes of the 30 Brand A batteries and the other displays the lifetimes of the 30 Brand B batteries presented in the worked example. (With window settings: $0 \leq x \leq 10$, $0 \leq y \leq 16$). Plot the histogram of the lifetimes of either brand using these settings and identify the matching histogram. Use class intervals of 1.0.

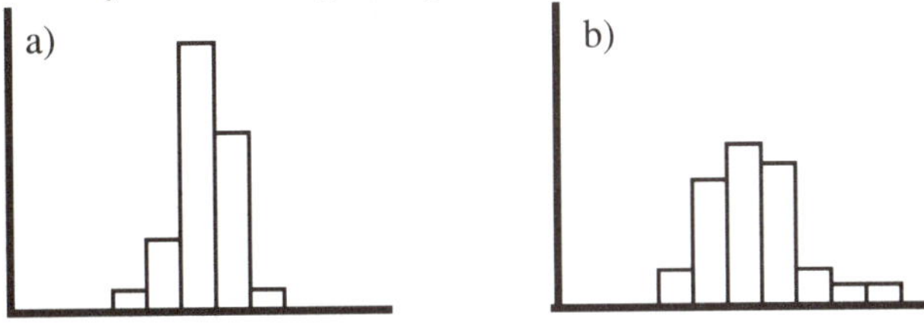

Double the class intervals in your histogram. What happens to the shape of the histogram when the class interval doubles? Why?

8. Describe how the shape of a histogram is related to the dispersion of a set of data.

9. Construct box plots for the diameters of the ball bearings manufactured by each of the two companies in exercise 6. Use the shapes of the box plots to determine which set of ball bearings has the smaller dispersion. Compare this conclusion with your answer to exercise 6.

10. Create (if possible) two sets of data which have the same mean, so that one of the sets has a greater standard deviation but a smaller interquartile range than the other set.

11. The vertical line with equation $x = \alpha$ divides the area of a histogram in half. Is α the value of the mean? the median?

12. A set of values z_i is derived from a set of measurements, x_i by the equation $z_i = \dfrac{x_i - \bar{x}}{\sigma}$ where σ is the standard deviation of the x_i. Calculate the mean and standard deviation for the z_i.

CHALLENGE

The symbol, Σ (upper case sigma) is used to denote a sum.

To write $x_1 + x_2 + x_3 + \ldots + x_n$ we write: $\displaystyle\sum_1^n x_i$

Write the definition of $\bar{x}$ and σ using the sigma notation.

Pressing **STAT** ► **ENTER** **ENTER** yields the display shown below.

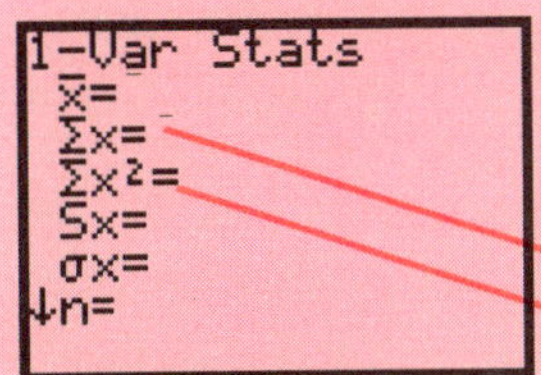

Enter the natural numbers from 1 to 10 into L_1 of the STAT edit table and use 1-Var Stats to find the sum of these numbers and the sum of their squares.

The students in Mr. Santos' class were asked to record the number of hours spent studying for their mathematics test. For each student, Mr. Santos wrote an ordered pair (x, y). The value of x represented the number of hours of study and the value of y was the student's mark on the test. Marsha's ordered pair was $(3, 82)$ because she spent 3 hours studying and received a final mark of 82 out of 100. The set of ordered pairs which Mr. Santos recorded are shown on the right.

Ordered Pairs

(3.0, 82)	(5.5, 78)	(1.0, 60)	(4.9, 93)	(5.1, 86)
(2.5, 71)	(4.2, 90)	(0.5, 40)	(3.5, 88)	(7.0, 96)
(1.5, 73)	(2.4, 82)	(2.0, 53)	(6.2, 87)	(8.4, 100)
(2.6, 75)	(3.7, 85)	(5.4, 70)	(9.3, 89)	(7.6, 87)
(1.4, 48)	(0.5, 56)	(6.5, 85)	(2.3, 61)	(5.2, 74)
(1.0, 47)	(3.5, 87)	(8.2, 94)	(3.0, 86)	(5.4, 92)

WORKED EXAMPLE 1

Use the data given above to determine whether there is a relationship between the number of hours of study and the test mark.

Solution

To enter these data, we access the STAT EDIT table by pressing: **STAT** **ENTER**
We clear the table and enter in L_1 the first coordinates of all the ordered pairs and in L_2, all the second coordinates, taking care to check that opposite each value in L_1 is the corresponding second coordinate in L_2.

The display shows the first 7 rows of the table.

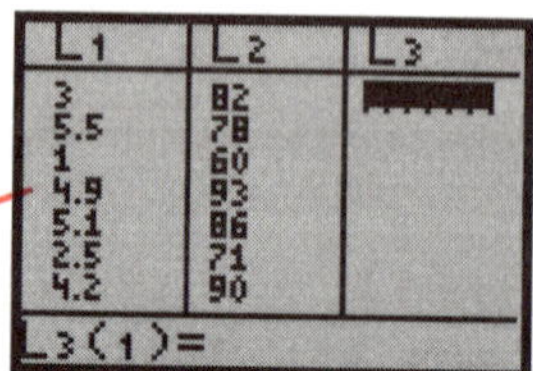

Row 4 corresponds to the ordered pair (4.9, 93)

To determine whether there is a relationship between the number of hours of study and the test mark we plot the 30 data points corresponding to the 30 ordered pairs. To do this, we first turn off all statistical plots by entering:

2nd [STAT PLOT] **4** **ENTER** followed by **2nd** [STAT PLOT] **1** **ENTER**

We define Plot 1 by selecting the options as shown in the display. That is, we choose the first icon in row "Type" which represents a type of plot called a *scatter plot*. We select L_1 as the Xlist and L_2 as the Ylist.

To graph this plot, we press: **ZOOM** **9** to get the display shown here.

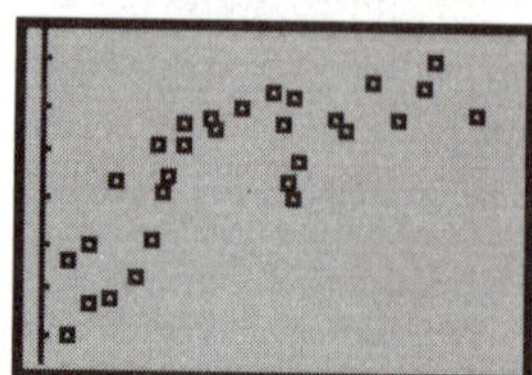

The bottom figure shows that almost all the data points in the scatter plot can be contained in an oval which is inclined along an axis with positive slope. In such a case we say that the variables plotted on both axes are *positively correlated*. The scatterplot shows that there is a positive correlation between preparation time for a test and the test mark. When the data points cluster close to the axis of the oval, we say the variables have a *strong positive correlation*. This scatterplot shows a fairly strong positive correlation.

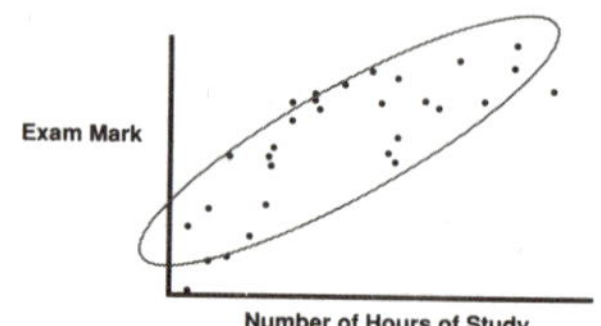

Exam Mark

Number of Hours of Study

IS A CAR AN INVESTMENT OR AN EXPENDITURE?

When an item which you purchase tends to increase in value, we usually refer to it as an *investment*. Conversely, an item which decreases in value, is called an *expenditure*. Understanding the difference between an investment and an expenditure is an important key to managing your personal finances. To determine whether a car is an expenditure or an investment, we must find out whether its value increases or decreases with its age.

The table displayed here shows for a popular model car, the resale value in dollars for each age between 1 and 10 years.

Age	Value	Age	Value
1	$15,975	6	$3,448
2	$ 9,285	7	$2,755
3	$ 8,000	8	$1,995
4	$ 6,790	9	$1,400
5	$ 3,150	10	$1,268

WORKED EXAMPLE 2

Use the data given in the table to determine whether there is a correlation between the age of a car and its resale value.

Solution

By scanning the table, we see that the resale value of the car tends to decrease with increasing age, so there seems to be a relationship, i.e. a *correlation*, between these two variables. To determine whether it is a strong or a weak correlation, we proceed as in worked example 1.

To enter these data, we access the STAT EDIT table by pressing: **STAT** **ENTER**
We clear the table and enter in L_1 the ages from 1 to 10 years and in L_2, all the corresponding resale values.

To determine whether there is a relationship between the age of that particular model car and the resale value, we plot the 10 data points as 10 ordered pairs. To do this, we first turn off all statistical plots by entering:

2nd [STAT PLOT] **4** **ENTER** followed by **2nd** [STAT PLOT] **1** **ENTER**

We define Plot 1 choosing the same options as in worked example 1, and press:
ZOOM **9** to get the display shown here.

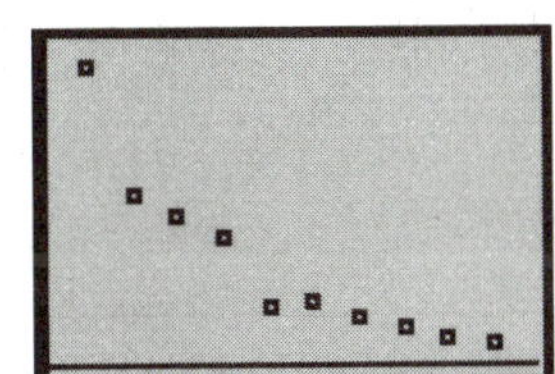

The display shows that almost all the data points in the scatter plot can be contained in an oval which is inclined along an axis with negative slope. In such a case we say that the variables plotted on both axes are *negatively correlated*. The scatterplot shows that there is a negative correlation between the age of a car and its resale value. This scatterplot shows a very strong negative correlation.

> **Note**: It is important to understand that if two variables are strongly correlated, it is not necessarily true that a change in one variable "causes" a change in the other. Often changes in both variables result from a common "cause". For example, shoe size and reading ability of school children are positively correlated because increases in both these variables come with increased age. However, having large feet does not "cause" increased reading ability.

1. Write in your own words the meanings of the following phrases:

- positive correlation
- negative correlation
- no correlation
- strong positive correlation
- strong negative correlation

2. Use the phrases above to identify the kind of correlation which is characterized by each of the following scatterplots.

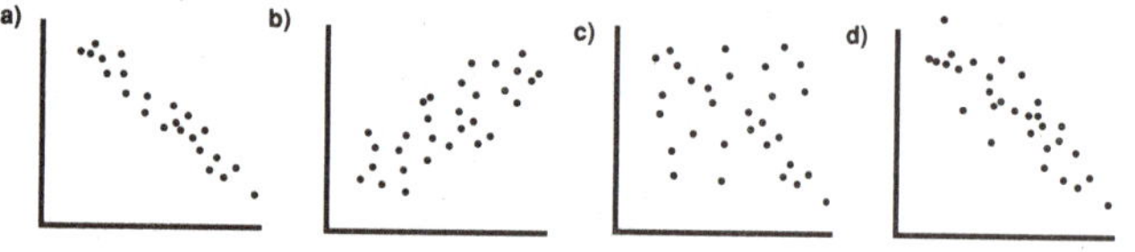

3. The table below shows 30 ordered pairs. The first coordinate in each ordered pair is the age of a randomly selected married woman. The second component is the age of her spouse. Use the procedure in the worked examples to create a scatterplot which will reveal whether there is a correlation between the age of a woman and her spouse.

Ages of Women and their Husbands				
(29, 34)	(37, 38)	(19, 20)	(57, 57)	(34, 32)
(23, 25)	(51, 54)	(72, 81)	(29, 23)	(70, 70)
(45, 54)	(39, 37)	(58, 56)	(64, 71)	(35, 35)
(42, 50)	(25, 24)	(37, 48)	(36, 36)	(27, 32)
(56, 42)	(17, 24)	(28, 28)	(57, 26)	(41, 39)
(47, 47)	(16, 18)	(24, 27)	(84, 87)	(55, 59)

If you discover a correlation, indicate whether it is positive or negative and describe it as strong or weak.

4. A biologist who was studying crickets, hypothesized that the number of chirps per minute made by a cricket is strongly correlated to the outside temperature increases. He recorded the data shown in the table. Create a scatterplot to test his hypothesis.

5. For each of the following pairs of variables, indicate what kind of correlation (if any) you would expect to exist. Explain why.
a) height and weight
b) age and personal savings
c) level of education and income
d) latitude of a U. S. city and its mean temperature in winter
e) A golfer's age and golf score
f) travel time and speed of travel for a particular distance
g) I. Q. and number of olives eaten per month

Temperature in °C	Chirps per min
17	105
18	110
19	110
20	126
21	126
22	130
23	130
24	152
24	156
25	160
26	170
27	171
28	175
29	196
30	212

6. In order to quantify the "degree" of correlation between two variables, Karl Pearson, one of the pioneers of statistics, defined the *correlation coefficient*, r between two variables x and y by the equation:

$$r = \frac{\sum_{i=1}^{n}(x_i - \overline{x})(y_i - \overline{y})}{\sqrt{\sum_{i=1}^{n}(x_i - \overline{x})^2 \sum_{i=1}^{n}(y_i - \overline{y})^2}}$$

where y_i is the value of y corresponding to $x = x_i$. $i = 1 \ldots n$

a) Suppose that x and y are so strongly correlated that y is a linear function of x; i.e. $y = ax + b$ for some constants a, b. Prove algebraically that $|r| = 1$. How are x and y related when $r = 1$? How are they related when $r = -1$?

b) What range of values of r would indicate: a strong correlation? a weak correlation? no correlation?

c) Enter into your STAT EDIT table, the temperatures and corresponding chirps per minute given in exercise 4. Use the STAT CALC SetUp menu to select the appropriate Xlist and Ylist. Then press: STAT ▶ 5 ENTER to find the correlation coefficient, r, for cricket chirps per minute and temperature.

7. a) Calculate the correlation coefficient between the ages of a woman and her spouse (See exercises 3 and 6.)
b) Calculate the correlation coefficient between hours of study and test mark presented in worked example 1.

DRAWING INFERENCES

A common error is the assumption that if two variables have a high correlation, then one of the variables is the "cause" of the other. This recent newspaper article revealed an interesting trend which resulted from such a false inference. Give a possible reason for the correlation.

Latin's Remarkable Resurrection

(New York)- A real resurrection of Latin is taking place. The upsurge is at the high school level and is very much a result of student demand…. Why are students flocking to Latin, and what do they expect to get out of it? …There is, allegedly, a clear correlation between taking Latin at school and doing well on scholastic aptitude exams, and the students are well aware, as they always are, of the statistics. They may not see why Latin, in the high-tech age should help them reach their goal, but they are clear on the fact that it does, and plausibly invoking the national adage, "If it works, don't fix it," queue up and sign on.

WILL FEMALES OUTRUN MALES IN THE 21ST CENTURY?

The women's 100-m dash was not an event in the Summer Olympic Games until 1928. In that year, Elizabeth Robinson of the U. S. won the event with a time of 12.2 seconds. Since that time, the women's record for the 100-m dash has decreased substantially.

The table below shows the winning times for the 100-m dash at the Summer Olympic games between 1928 and 1992 for both men and women. Perusal of the table reveals that the winning times seem to be decreasing more quickly for women than for men. Will women eventually outrun the men?

Wilma Rudolf of the United States wins the 200-m dash by 4 meters. Her victory in the 100-m dash made her the first American woman to win both events.

Year	Men's Record (in seconds)		Women's Record (in seconds)	
1928	Percy Williams (Canada)	10.80	Elizabeth Robinson (U.S.)	12.20
1932	Eddie Tolan (U. S.)	10.30	Stanislawa Walasiewicz (Poland)	11.90
1936	Jessie Owens (U. S.)	10.30	Helen Stephens (U. S.)	11.50
1948	Harrison Dillard (U. S.)	10.30	Francina Blanker-Koen (Nethlds)	11.90
1952	Lindy J. Remigino (U. S.)	10.40	Marjorie Jackson (Australia)	11.50
1956	Bobby J. Morrow (U. S.)	10.50	Betty Cuthbert (Australia)	11.50
1960	Armin Hary (Germany)	10.20	Wilma Rudolf (U. S.)	11.00
1964	Robert L. Hayes (U. S.)	10.00	Wyomia Tyus (U. S.)	11.40
1968	James Hines (U. S.)	9.90	Wyomiz Tyus (U. S.)	11.00
1972	Valery Borzov (U.S.S.R.)	10.14	Renate Stecher (East Germany)	11.07
1976	Hasely Crawford (Trin. & Tob.)	10.06	Annegret Richter (West Germany)	11.01
1980	Allan Wells (Great Britain)	10.25	Lyudmila Kondratyeva (U.S.S.R.)	11.06
1984	Carl Lewis (U. S.)	9.99	Evelyn Ashford (U. S.)	10.97
1988	Carl Lewis (U. S.)	9.92	Florence Griffith-Joyner (U. S.)	10.54
1992	Linford Christie (Great Britain)	9.96	Gail Devers (U. S.)	10.82

A set of measurements, such as winning times in the 100-m dash, taken at regular intervals of time is called a *time series*. By plotting these measurements against time and joining the plotted points with line segments we create an *xyLine*. This xyLine enables us to investigate the data for upward or downward trends. In the worked example on page 26 we construct xyLines to investigate the winning times in the 100-m dash for possible trends. We then construct for the women's winning times and the men's winning times, a "trend line" called a *median-median line* which smooths out short term fluctuations.

WORKED EXAMPLE

a) Use the data given in the table on page 25 to construct the xyLines for the Olympic winning times in the 100-m dash for men and for women during the period from 1928 to 1992.

b) Construct the median-median line for the men's and for the women's winning times and trace along these lines to find the point of intersection.

Solution

a) We enter the values from the table into the STAT EDIT table to obtain the table shown in the display, in which L_1 denotes the year, L_2 the men's record time and L_3 the women's record time.

L1	L2	L3
1928	10.8	12.2
1932	10.3	11.9
1936	10.3	11.5
1948	10.3	11.9
1952	10.4	11.5
1956	10.5	11.5
1960	10.2	11
1964	10	11.4
1968	9.9	11
1972	10.14	11.07
1976	10.06	11.01
1980	10.25	11.06
1984	9.99	10.97
1988	9.92	10.54
1992	9.96	10.82

L1(16)=

We define Plots 1 and 2 as shown on the displays. Note that we select the xyLine (second icon in "Type") for both plots and different "Marks" for the two plots.

When we activate ZoomStat by pressing: [ZOOM] [9] , we obtain this display.

The top xyLine in the display is Plot2, representing the women's winning times. This xyLine seems to be sloped downward more steeply than the winning times for men.

b) To calculate the median-median line for the men's winning times, we press:

 [STAT] [▶] [3] to access the STAT SetUp menu.

We then select L_1 for the Xlist and L_2 for the Ylist under the row titled **2-Var Stats**. Then to calculate the median-median line, we press: [STAT] [▶] [4] [ENTER]
We obtain equation of the median-median line shown in the display.

To plot this equation, we must enter it onto the [Y =] list by pressing:

      [Y =] [VARS] [5] [▶] [▶] [7] [ENTER]

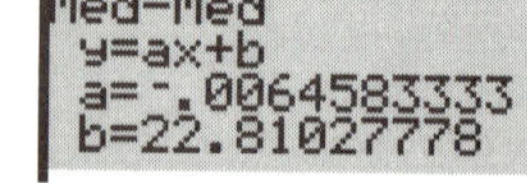

We enter the median-median line for women's winning times onto the [Y =] list by following the same procedure except we enter L_3 for the Ylist when setting up the **2-Var Stats.** The equation of the women's median-median line is shown in the display.

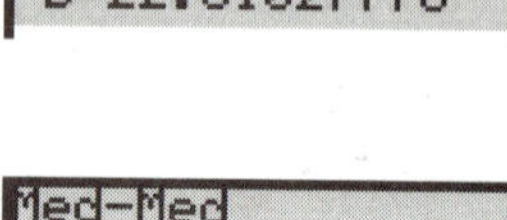

To graph the xyLines together with the median-median lines, we ensure that Plot1 and Plot2 are turned "on" and that the equations of both median lines are selected on the [Y =] list and then press: [ZOOM] [9] . We get the display shown here.

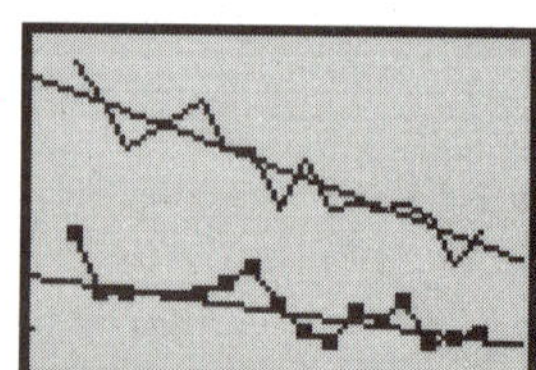

It is clear that the median-median line for the women's winning times has a steeper negative slope than the median-median line for the men's winning times and therefore the two lines will intersect at some point. Since the lines do not intesect in the current window, we redefine the window settings so that $X_{max} = 2100$ and $Y_{min} = 9$.

To find the point of intersection, we press: [GRAPH] [2nd] CALC [5]

In response to the prompts, **First curve?** and **Second curve?**, we press [ENTER]

In response to the prompt, **Guess?**, we move the cursor close to the intersection and press [ENTER]. We obtain this display which suggests that if women and men continue to improve at present rates, around the year 2052, the winning time for both genders will be about 9.56 seconds. According to this model, what happens after 2052?

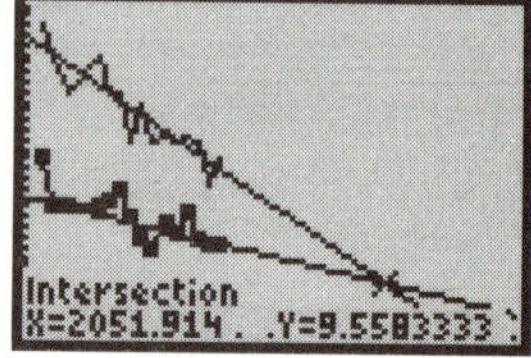

1. Write a sentence or two to explain each of the following terms.
- time series
- xyLine
- trend line
- median-median line

2. Do each part of this exercise in sequence to construct a median-median line without using your calculator.

a) Use graph paper to create a grid like the one below.

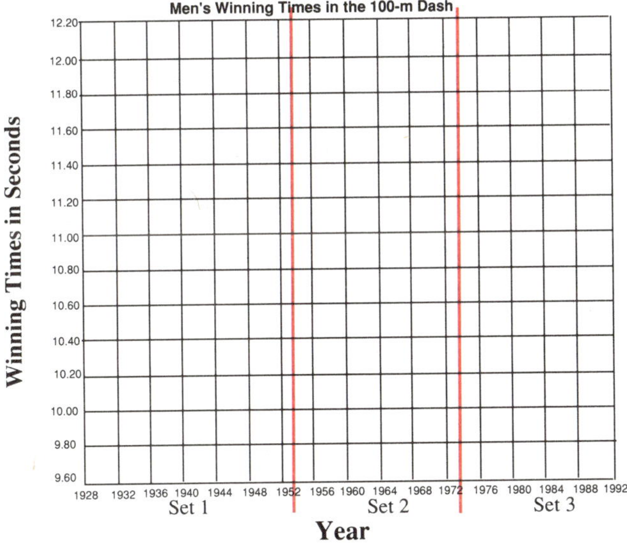

b) Using the table on page 25, plot the points corresponding to the men's winning times.

c) Partition the data into three sets as shown so that there are five consecutive Olympic years in each set.

d) Identify the median year in set 1 and denote it by x_1, and the median winning time in set 1 and denote it by y_1. Form the ordered pair (x_1, y_1) and call it the *summary point* for set 1.

e) Record the summary points (x_2, y_2) and (x_3, y_3) for set 2 and set 3 respectively.

f) Draw a line through (x_1, y_1) and (x_3, y_3). Draw a second line parallel to the first and through (x_2, y_2). Finally draw a third line between and parallel to the first two so that the distance between the first and third lines is one-third the distance between the first and second lines. This line is the *median-median line*.

g) Measure the slope of the median-median line and its y-intercept and then compare with the equation of the median line for the men's winning times in the worked example.

3. Some scientists believe that increased quantities of carbon dioxide in the earth's atmosphere are causing a gradual global warming that will eventually melt the polar ice caps and cause world-wide flooding. The table below shows the annual number of metric tons of carbon dioxide emissions per capita from fossil fuel consumption and cement manufacture during the period 1950-1989.

Per Capita Carbon Dioxide Emissions in Metric Tons for the period 1950-1989

Year	Emissions	Year	Emissions
1950	2.38	1975	4.14
1955	2.71	1980	4.32
1960	3.15	1985	4.07
1965	3.48	1989	4.21
1970	4.07		

a) Follow the procedures outlined in exercise 2 to construct the median-median line for the annual number of tons of carbon dioxide emissions per person between 1960 and 1989.

b) Use your TI-82 calculator to calculate the median-median and compare with your answer in part a).

c) If the carbon dioxide emissions continue to increase as in the past, what will be the per capita emissions of carbon dioxide from fossil fuels and cement manufacture in the year 2050?

d) What assumptions do we make when we construct a median-median line and use it to predict the values of a variable at some future date? For what reasons might such assumptions not be valid?

DRAWING INFERENCES

The median-median lines in the worked example suggest that the winning times for females and males in the 100-m dash will be about the same around the year 2057. Do you believe this prediction? Give reasons for your answer.

Use your median line to predict the winning time for women in the 100-m dash in the year 2150. Do you believe this prediction? Explain your answer.

Karl Friedrich Gauss is regarded by historians of mathematics to have been a mathematician without peer (with the possible exceptions of Archimedes and Newton). The list of branches of mathematics which he developed or extended is too large for inclusion here. Suffice it to say that his genius for mathematics manifested itself when he was three years old and remained with him until his death in his 78th year.

When Gauss was 18 years old, he was developing a theory of errors in measurement. Out of this work, the ubiquitous *normal distribution* emerged (see Exploration 14). In exploring the question as to which of a large number of measurements is most likely the correct one, Gauss invented the *method of least squares*.

We learned in Exploration 6, that when two variables are strongly correlated, the points in their scatterplot cluster along a straight line. Of all possible straight lines, which one fits the data the best? To find the equation of this optimum line which we call the *line of best fit*, Gauss computed the values of the constants "a" and "b" in the equation:

$$y = a + bx.$$

such that the sum of the squares of the vertical distances of the data points from the line with equation, $y = a + bx$ is a minimum. (See graph on p29.) The vertical distance of each data point, (x_i, y_i) from the line is called a *residual*. Hence the line of best fit is merely the straight line for which the sum of the squares of the residuals is a minimum.

The importance of the line of best fit is that it enables us to predict the value of the y variable when we know or can measure the value of the x variable, and conversely (provided the two variables are strongly correlated). In 1855, Sir Francis Galton investigating inherited traits, constructed a scatter plot of the heights of 928 adults against those of their parents. He discovered that these variables are strongly correlated. Furthermore, children of tall (short) parents were taller (shorter) than average but closer to the average than their parents. That is, extremes in height "regressed" closer to the mean in successive generations. Hence the line of best fit relating these two variables was called the *regression line*.

Often the variables, x and y are related in a non-linear way. In such a case, we seek a regression curve which is not a straight line. For example, the quadratic regression curve is the parabola of best fit; that is the parabola with equation $y = ax^2 + bx + c$, where a, b and c are chosen so that the sum of the squares of the residuals is a minimum. The equations (which define the regression curves) which you can access on your TI-82 are listed on your STAT CALC menu as shown here.

Linear regression curve	equation:	$y = ax + b$
Quadratic regression curve	equation:	$y = ax^2 + bx + c$
Cubic regression curve	equation:	$y = ax^3 + bx^2 + cx + d$
Quartic regression curve	equation:	$y = ax^4 + bx^3 + cx^2 + dx + e$
Linear regression curve	equation:	$y = a + bx$
Logarithmic regression curve	equation:	$y = a + b\ln x$
Exponential regression curve	equation:	$y = ab^x$
Power regression curve	equation:	$y = ax^b$

```
EDIT CALC
5↑LinReg(ax+b)
6:QuadReg
7:CubicReg
8:QuartReg
9:LinReg(a+bx)
0:LnReg
A:ExpReg
B:PwrReg
```

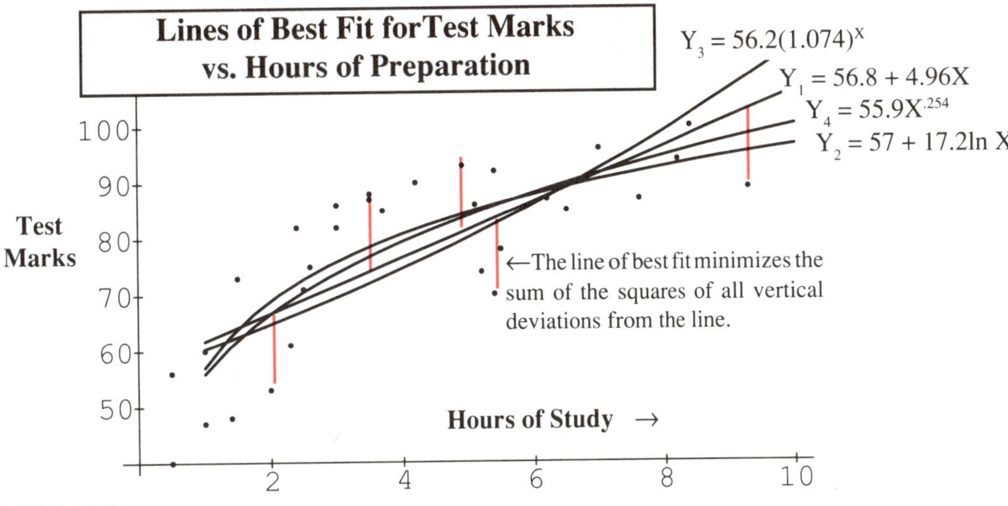

WORKED EXAMPLE

The graph above shows a scatter plot of the data on page 22, relating the marks on a mathematics test with the number of hours of study preparation.

a) Use the data on page 22 to create a scatter plot like the one above on your TI-82.
b) Find the equations of the linear and the logarithmic regression lines.
c) Graph the regression lines so they are superimposed on the scatter plot.
d) Which of the regression curves yields the better fit as defined by Gauss?
   Use that line to estimate the mark a student would achieve with 3.2 hours of study.

Solution

a) We follow the procedure in worked example 1 on page 22 to enter the ordered pairs into the STAT EDIT table and to graph the scatterplot.

b) To calculate the equation of the regression line (with equation of the form $Y = a + bX$) we enter: **STAT** **▶** **3** to access the STAT SetUp menu.

We then select L_1 for the Xlist and L_2 for the Ylist under the row titled **2-Var Stats**. Then to calculate the regression line, we press: **STAT** **▶** **5** **ENTER**

We obtain equation of the regression line shown in the display. ————

We define this equation as Y_1 on the **Y =** list by pressing:

Y = **VARS** **5** **▶** **▶** **7** **ENTER**

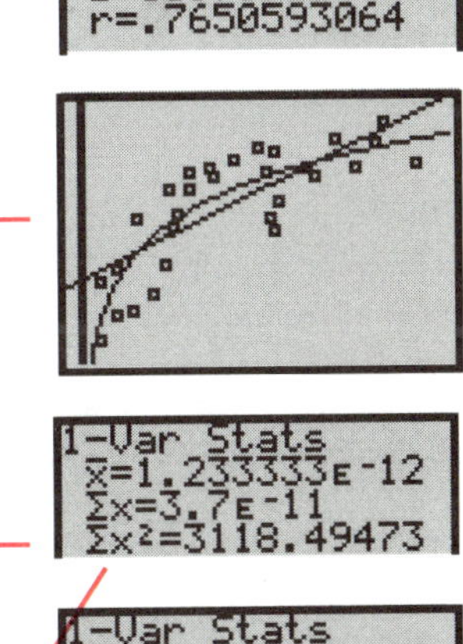

We proceed as above to obtain the equation of the logarithmic regression curve.

c) When we press: **ZOOM** **9** we get the display shown here. ————

d) To determine which curve is the better fit for the data, we calculate the residuals; that is the difference between the observed value (given data) and the predicted value (y coordinate on the regression line). To calculate the list of residuals L_4 for the linear regression, we access the STAT EDIT table and define: $L_3 = Y_1(L_1)$ and $L_4 = L_3 - L_2$.
We evaluate the sums of the squares of the residuals by evaluating **1-Var Stats** with the Xlist as L_4 and the frequency as 1. This yields the display shown here. ————

To calculate the list of residuals L_6 for the logarithmic regression curve, we define: $L_5 = Y_2(L_1)$ and $L_6 = L_5 - L_2$ and proceed as above to evaluate the sums of the squares of the residuals. Since the sum of squares of the residuals is less for the logarithmic curve, it provides the better fit. We then store 3.2 in variable, X, and evaluate Y_2 to get an estimated mark of about 77.

1. Write a sentence or two to explain the meaning of each of the following terms.
- method of least squares
- regression line
- regression curve
- line of best fit
- residual

2. For each of the following scatterplots estimate the slope of the straight line of best fit. (Assume the x and y-scales are identical.)

3. Explain how a median-median line and a regression line are alike and how they are different.

4. Write the general equations of the curves of best fit (using "a" and "b" as parameters) in each of the following: (The first one is done for you.)
- the linear regression model $Y = a + bX$
- the logarithmic regression model
- the exponential regression model
- the power regression model

5. Explain how you can determine which regression curve of several different types fits a given set of data best.

6. In the worked example, we compared the regression line and the logarithmic regression curve in fitting the data pertaining to hours of study and test marks.

a) Would you expect the relationship between hours of study and test marks to have been linear? Give reasons for your answer.

b) Why might you expect the relationship between hours of study and exam mark to be logarithmic?

7. a) Using the table on page 25, enter the Olympic year in column L_1 of your STAT EDIT table and the men's winning times in column L_2. Then graph a scatter plot.

b) Calculate the linear and logarithmic regression equations. Which regression curve, the linear or the logarithmic, provides a better match to the data? Why?

c) Use the regression curve of better fit to estimate the men's winning time at the Summer Olympics in 2052. Compare this estimate with the median-median line estimate on page 26. Explain reasons for any difference in the two estimates.

8. Using the data in exercise 3 on page 24, construct a scatterplot for the ages of women and their husbands.

a) What is the correlation between the age of a woman and her spouse? Is this a strong correlation? Calculate and graph the equation of the regression line for this scatter plot.

b) Graph the logarithmic regression curve. Which curve, the linear or the logarithmic, provides a better fit to the data? Why do you think this curve provides a better fit?

c) Use the better regression curve to estimate the age of a man who is married to a 48-year-old woman.

9. The ordered pairs in the table below show the U.S. population at the beginning of each decade from 1800 to the present. (The first coordinate gives the year and the second coordinate the population in millions.)

(1800, 5.31)	(1810, 7.24)	(1820, 9.64)	(1830, 12.9)
(1840, 17.1)	(1850, 23.2)	(1860, 31.4)	(1870, 39.8)
(1880, 50.2)	(1890, 62.9)	(1900, 76.0)	(1910, 92.0)
(1920, 105.7)	(1930, 122.8)	(1940, 131.7)	(1950, 150.7)
(1960, 179.3)	(1970, 203.3)	(1980, 226.5)	(1990, 248.7)

Make a scatterplot of the U.S. population vs the date. Which regression curve yields the best fit to this data? Use it to estimate the U. S. population in 1965 and in 2010.

10. The table below shows the ages and the systolic blood pressures of 10 women.

Age vs. Systolic Blood Pressure for a Sample of 10 Women

Age	Blood Pressure	Age	Blood Pressure
34	120	55	140
38	114	57	152
44	140	60	157
44	136	63	149
49	147	72	162

Graph a regression curve that is a best fit of the data. Use your regression curve to estimate the blood pressure of someone who is 80 years of age.

CHALLENGE

Collect 20 measurements of any two variables which you believe may be correlated. Construct a scatter plot of your data. Record the correlation coefficient of the two variables. If the two variables are correlated, construct a curve of best fit and use it to estimate the dependent variable for a given value of the independent variable.

Introduction to Probability

© 1994 by Sidney Harris

1 Humans have gambled for money from ancient times. However, it was not suspected until the late fifteenth century, that chance and "luck" might actually be subject to mathematical patterns. Girolamo Cardano an ardent gambler, physician and mathematician, in 1539 published a gambler's manual reporting patterns in the outcomes associated with the toss of a die.

2 In 1657, Christiaan Huygens, the Dutch physicist published the first formal treatment of Probability Theory, titled, *On Reasoning in Games of Chance*.

AN IMPORTANT COUNTING PRINCIPLE

The computation of theoretical probabilities is based upon an important counting principle called *logical multiplication*.

Simply stated, if a person has 3 different sweatshirts and 2 pairs of slacks, then there are 3×2 or 6 possible outfits which can be composed. We can illustrate this with a *tree diagram*.

One sweatshirt carries the insignia *Bulls* while another displays *Bears* and the third, *Hawks*.

One pair of slacks is blue, the other is striped.

The tree diagram shows that for each selection of a sweatshirt, there are two possible selections of slacks.

Shirts	Slacks
Bulls	Blue / Striped
Bears	Blue / Striped
Hawks	Blue / Striped

In general:

> **The Fundamental Counting Principle**
> If one item can be selected in *m* ways, and for each of these selections a second item can be selected in *n* ways, then there are *mn* ways of selecting the two items.

3 In a collabora... the Problem of the Points... foundations of Probability...

Roberval argued that the possible outcomes are:

First Toss	Second Toss	Result
Heads	Game over	A wins
Tails	Heads	A wins
Tails	Tails	B wins

The probability that A will win is $\frac{2}{3}$.

The Probl...

A and B play...

A fair coin i...
If a head oc...
then A wins.

What is the pr...

THE PERMUTATION CONCEPT

In how many ways can we arrange these letters?

C A T

Using a tree diagram, we see there are 6 arrangements.

That is, we can choose the first letter in 3 ways.
For each choice of a first letter, there are 2 ways to choose the second letter.
The third letter can be chosen in only one way.

The number of arrangements or *permutations* is 3×2×1 or 3! (pronounced "3 factorial".)

C — A–T / T–A
A — C–T / T–C
T — A–C / C–A

In general, the number of arrangements or permutations of n different objects is n×(n-1)×...3×2×1 or n! ("n factorial").

Pierre de Fermat
1601-1665

THE BETTMANN ARCHIVE

④ The book, *Ars Conjectandi*, written by Jakob Bernoulli and published posthumously in 1713, was the first authoritative treatment of Probability Theory.

Jakob Bernoulli
1654-1705

THE BETTMANN ARCHIVE

The most important questions of life are, for the most part, really only problems of probability. ... so that the entire system of human knowledge is connected with this theory...It is remarkable that probability, which began with the consideration of games of chance, should have become the most important object of human knowledge.
—Pierre Simon de Laplace

Pierre Simon de Laplace
1749-1827

...mulated by ...l and Fermat developed the ...y in the early to mid 1650's.

the Points

...lowing game:

...d twice. ...either toss, ...wise, **B** wins.

...y that **A** wins?

Pascal argued that the possible outcomes are:

First Toss	Second Toss	Result
Heads	Tails	A wins
Heads	Heads	A wins
Tails	Heads	A wins
Tails	Tails	B wins

The probability that A will win is $\frac{3}{4}$.

APPLICATIONS TO PROBABILITY

WORKED EXAMPLE

What is the probability that a random selection of 3 names from a field of 10 horses will yield the winner, the second place and the third place horses in the order they were selected?

Solution

We must compute the probability of a particular permutation of 3 names out of *all possible permutations of 3 names chosen from 10 names*. The number defined in italics is denoted $_{10}P_3$. In general, $_nP_r$ is the number of permutations of **n** objects taken **r** at a time. The probability of an event is defined by:

$$\text{Probability of an event} = \frac{\text{Number of outcomes favorable to the event}}{\text{Number of outcomes possible}}$$

Probability of choosing the first 3 horses in the correct order $= \dfrac{1}{_{10}P_3}$ ← favorable permutations ← possible permutations

To evaluate $_{10}P_3$ we press: **10** **MATH** **◄** **2** **3** **ENTER**

We obtain, $_{10}P_3 = 720$, so the probability is 1 chance in 720. To express this as a decimal fraction, we press **x^{-1}** **ENTER** to obtain $.001\overline{38}$

EXERCISES

1. Use your calculator to evaluate the following.

   a) $_4P_1$ b) $_4P_2$ c) $_4P_3$ d) $_4P_4$
   e) $_5P_1$ f) $_5P_2$ g) $_5P_3$ h) $_5P_4$

2. a) Use your answers to exercise 1 to conjecture a formula for $_nP_r$. (**Hint:** Consider quotients of factorials and define 0! to have the value 1.)
b) Use a counting argument to prove your formula for $_nP_r$, where $_nP_r$ denotes the number of permutations of n objects taken r at a time and r ≤ n.
c) Verify your formula for $_nP_r$ reduces to n! when n = r.

3. How many different 6-letter licence plates can be made using only the letters of the alphabet if no letter can be used more than once on any plate? Write your answer using the notation $_nP_r$, then compute this number.

4. If security codes are composed of 4 different digits, what is the probability that a security code issued to you would be the year of your birth?

5. In a triactor race, the person placing the wager must pick *in the correct order* the horses which finish first, second and third. What is the probability of winning the triactor in a field of 9 horses if the horses are selected randomly?

Father Gregor Mendel
1822-1884

In 1850, Father Gregor Mendel failed his teacher certification examinations. His lowest mark was in biology. Though uncertified, he was permitted to teach natural science at the technical high school in Brünn, Austria. There, in the little monastery garden, he conducted experiments with pea plants. He reported the results of these experiments in 1865 to the Natural Science Society at Brünn as, "experiments in plant hybridization". One of his observations was that "hybrids [when "mated"] form seeds having one or the other of two differentiating characteristics". This hypothesis laid the foundations for the modern theory of genetics!

A Short Summary of Mendel's Experiments

Mendel conducted his series of experiments between 1856 and 1864, while Charles Darwin was simultaneously formulating his theory of natural selection. Mendel compared pea plants with respect to 7 different pairs of characteristics such as: tall vs. short, green seeds vs. yellow seeds and wrinkled seeds vs. smooth seeds. In the" tall vs. short" experiment, he began with two kinds of *pure* plants; i.e. those which came only from generations of tall plants and those which came only from generations of short plants. When he "mated" *pure* tall plants with *pure* short plants, the offspring were always tall. When he mated any pair of tall offspring he discovered that about one in four of these second generation plants were short. The table below summarizes some of his results.

EXPERIMENT 1	EXPERIMENT 2
Procedure: He cross-pollinated pure tall pea plants with pure short pea plants.	**Procedure:** He cross-pollinated the hybrids formed in experiment 1 and grew their offspring.
Observation: All the first generation plants from these "pure" parents were tall.	**Observation:** One quarter of these second generation plants were short and the rest were tall.
Interpretation: When pure tall plants "mate" with pure short plants, all the offspring are tall. The characteristic of "shortness" disappears in the first generation.	**Interpretation:** When hybrid plants "mate", the characteristic which disappeared in the first generation, returns to one quarter of the offspring in the second generation.

Experiment 1
Mating pure tall with pure short yields all tall offspring

Experiment 2
When tall offspring are mated, one in four is short.

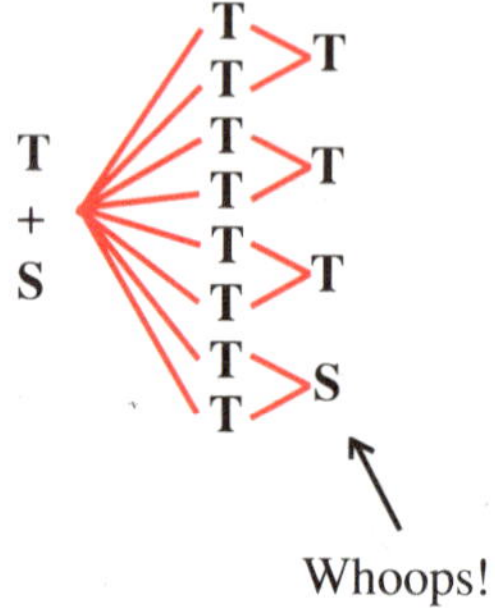

Mendel's Conclusions

1. Since the offspring are either tall or short (not medium), such characteristics are determined by a particular "gene".

    A "pure" tall pea plant carries two T genes, denoted {T, T}
    A "pure" short pea plant carries two s genes, denoted {s, s}.

2. Each offspring receives one gene from each parent.

3. If an offspring receives at least one T gene, it will be tall.
    That is, the T gene is *dominant*

Using Probability Theory to Explain Mendel's Results

Mendel used his three conclusions (see page 34) and a probability model to explain the results of his experiments as follows:

Experiment 1

The four different ways of drawing a height gene from a "pure" tall parent and a height gene from a "pure" short parent are shown in the table.

In all cases, the offspring inherits one T and one s, and since T is dominant the offspring is tall.

		Tall Parent	
		T	T
Short	s	sT	sT
Parent	s	sT	sT

Experiment 2

The four different ways of drawing a height gene from two hybrid parents are shown in the table.

In three of the cases, the offspring has a T gene, and since T is dominant, the offspring will be tall. In the fourth case, there is no T gene, so the off-spring will be short.

		Hybrid Parent	
		s	T
Hybrid	s	ss	sT
Parent	T	sT	TT

Using this probability model, Mendel was able to explain his observations and establish some fundamental laws of heredity. He published these results in 1866. However, the significance of his discoveries was not realized until 16 years after Mendel's death, when three European biologists independently discovered some of these principles. The table above which explains why three-quarters of the "grandchildren" inherit the dominant characteristic of the "grandparent" is exactly the table which explains the solution to the Problem of the Points — the very problem which launched the development of Probability Theory! (See pp 32-33). It is interesting that a branch of mathematics originating in games of chance has found widespread use in almost every facet of life.

The mathematics of heredity involves counting the various *combinations* of outcomes in the random selection of genes from two parents. We note that it is the *set* of genes which is important and order of selection does not play a role. That is, {s, T} is equivalent to {T, s}. The fact that about 6 billion people live on earth and that no two are identical (including identical twins) is a testimonial to the large number of combinations possible when selecting a set of genes from a large number of possible genes. In the worked examples, we demonstrate how to count the number of subsets or *combinations* of a given set of elements.

Definition: The selection of elements from a set is called a *combination*. The number of combinations of n elements taken r at a time is denoted $_nC_r$ (read *n choose r*.)

For example, from the set of 5 elements, {a, b, c, d, e} there are 5 combinations taken 4 at a time, namely:
{a, b, c, d}, {a, b, c, e}, {a, b, d, e}, {a, c, d, e}, {b, c, d, e}.

A combination is independent of the order in which its elements are listed.

WORKED EXAMPLES

WORKED EXAMPLE 1

Find a formula in terms of n and r for the number of combinations of n objects taken r at a time. Use your formula to find the number of subcommittees of 5 people which can be selected from a committee of 20 people.

Solution

Let $\{x_1, x_2, x_3, \ldots x_r\}$ denote a combination of r elements from a set of n elements.
For each such combination, we can arrange the r elements in r! (r factorial) ways to form r! permutations.

That is, the number of permutations of n objects taken r at a time is r! times the number of combinations. Hence,

$$_nP_r = r! \, _nC_r$$

Number of permutations of n objects taken r at a time	Number of combinations of n objects taken r at a time

Solving for $_nC_r$, we obtain: $\quad _nC_r = \dfrac{_nP_r}{r!} \quad$ or $\quad \dfrac{n!}{(n-r)!\,r!}$

To find the number of subcommittees of 5 from a committee of 20, we must evaluate $_{20}C_5$. To do this on the calculator, we press:

 **20** MATH ◄ **3 5** ENTER to obtain 15 504

WORKED EXAMPLE 2

If both genders are equally likely at birth, what is the probability that exactly 3 of the 5 children will be girls?

Solution

One possible configuration of 3 girls would be GGBBG; that is, the first two births are girls, followed by two boys and a girl. To count the number of such configurations, we count the number of ways of selecting 3 positions which are to be occupied by girls out of the 5 positions available. This is $_5C_3$.

The number of possible configurations is 2^5, since each of the 5 positions can be filled in 2 ways, B or G.
Therefore the probability of exactly 3 girls is $\dfrac{_5C_3}{2^5}$ or $\dfrac{5}{16}$.
The red boxes in the tree diagram show the 10 configurations out of 32 which have exactly 3 girls.

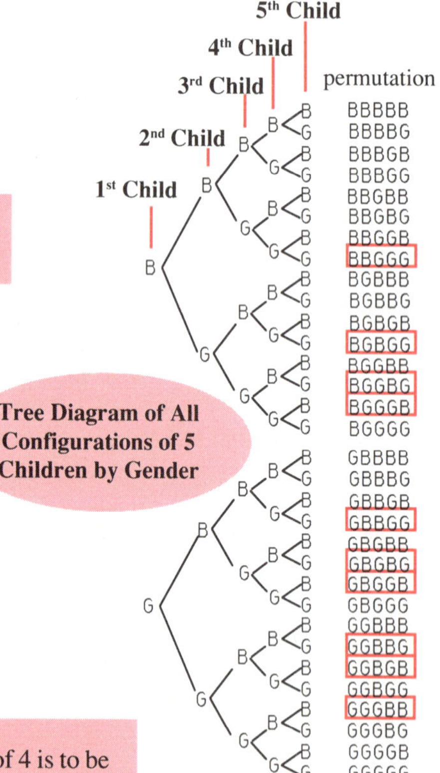

Tree Diagram of All Configurations of 5 Children by Gender

WORKED EXAMPLE 3

A government committee has 7 Republicans and 5 Democrats. A subcommittee of 4 is to be selected by lot. What is the probability that the subcommittee has exactly 2 Democrats?

Solution

There are $_{12}C_4$ ways of selecting a subcommittee of 4 from 12.
There are $_5C_2$ ways of selecting 2 Democrats from 5 of them, and for each of these, $_7C_2$ ways of selecting 2 Republicans.
Therefore, the probability of selecting a subcommittee of 4 containing exactly 2 Democrats is: $\dfrac{_5C_2 \times _7C_2}{_{12}C_4}$ or $\dfrac{42}{99}$

1. Use your calculator to evaluate the following.

   a) $_8C_4$ b) $_{12}C_{10}$ c) $_{21}C_{15}$ d) $(_{12}C_{10})^{-1}$

2. Use your calculator to verify that $_nP_r = r!\,_nC_r$ for

   a) $n = 9$, $r = 4$ b) $n = 17$, $r = 12$

3. How many different hands of 5 cards can be dealt from a deck of 52 cards?

4. a) An association of 20 people is to elect an executive committee of 3 members. In how many different ways can this committee be chosen?

   b) Once the three members have been chosen, in how many different ways can these members be assigned to the positions of president, vice president and treasurer?

   c) Use your answers to parts a) and b) to determine the number of ways the positions of president, vice president and treasurer of an association of 20 people can be assigned.

5. How many different 7-digit telephone numbers can be issued if the first 3 digits must be 335, 363 or 339?

6. The United States and Canada are among 14 nations which have entered a team in a world hockey tournament. To determine which teams will play in the first game, the names of the countries are thrown into a hat and two names are drawn. What is the probability that names of the United States and Canada will be drawn?

7. a) Calculate the probability that a family of 4 children will have an equal number of girls and boys if both genders are equally likely?

   b) Draw a tree diagram to verify your answer.

8. If both genders are equally probable at birth, what is the probability that there are exactly 4 boys in a family of 7 children.

9. a) What is the probability that a student could randomly choose exactly 7 correct answers on a 10-item true-false test?

   b) What is the probability that a student could guess *at least* 7 correct answers on the test?

10. a) In how many ways can a dance committee of 12 students be chosen from 9 boys and 8 girls if:

     i) there are no restrictions?

     ii) the committee must have exactly 6 boys?

     iii) the committee must have at least 6 girls?

   b) If the dance committee is chosen without restriction, what is the probability that it contains exactly 6 boys?

Exercises

11. In a triactor race, the person placing the wager may choose to *box* 3 horses. That is, the person selects 3 horses and if they come in first, second and third, *in any order*, the person wins. What is the probability of winning such a *box* bet when 3 randomly selected horses are randomly selected from a field of 9 horses? 

12. If security codes are composed of 4 different digits, what is the probability that a security code issued to you would be the year of your birth? (Use a different method from the one you used in exercise 5 of Exploration 9.)

13. Any line segment which joins two non-adjacent sides of a polygon is called a *diagonal* of the polygon. The red line segment in the diagram is a diagonal of the octagon.

     a) How many diagonals has a pentagon?

     b) How many diagonals has a hexagon?

     c) How many diagonals has an octagon?

Write an expression for the number of diagonals of a polygon of n sides.

14. There were **n** representatives at the annual meeting of theoretical mathematicians and each representative shook hands with all the other representatives. How many handshakes took place?

15. Ten thank you cards (each with a personal message) are placed randomly into ten addressed envelopes. What is the probability that exactly 5 of the cards are placed in the correct envelopes?

16. Prove that $_nC_r = {_nC_{n-r}}$.

   a) algebraically   b) by a combinatorial argument

17. Prove Pascal's rule algebraically:

$$_nC_r = {_{n-1}C_{r-1}} + {_{n-1}C_r}$$

CHALLENGE

In Mendel's Pea experiments, all of the first generation plants were tall. Three quarters of the second generation plants, produced by "mating" two hybrids, were tall.

Suppose these second generation plants were then "mated" to produce third generation plants. What fraction of the third generation plants would be tall?

In Explorations 9 and 10, we observed that theoretical probability is based on the idea that certain outcomes are equally likely to occur. If an experiment has n equally likely possible outcomes, we can assign a probability of $\frac{1}{n}$ to each outcome.

The theoretical probability of more complex outcomes or *events* can then be expressed in terms of the probabilities of these simple outcomes. However, in some experiments, such as the tossing of a paper cup, (See exercise 2.) the probabilities of the possible outcomes are not known and must be estimated by experiment. To provide a basis for the concept of *experimental probability*, we define *relative frequency*.

Definition: If an outcome A occurs f times on n trials of an experiment, then:

the *relative frequency* of outcome A $= \dfrac{f}{n}$

The cartoon suggests that the concept of "fair" may be unrealizable in the real world. In the worked examples, we investigate the relative frequencies of the outcomes in the toss of a fair coin as the number of tosses is increased.

Using the concept of relative frequency, we can define the *experimental probability* of an outcome as follows:

Definition: The *experimental probability* of an outcome A is the relative frequency of A for a large number of trials.
In more formal notation, we would write:

$$experimental\ probability\ of\ \text{A} = \lim_{n \to \infty} \frac{f}{n}$$

(This definition is not rigorous, since a formal limit does not exist.)

A *fair* coin is defined to be a coin for which the heads and tails outcomes are equally likely. Time constraints make it impractical to conduct experiments involving a large number of coin tosses. To *simulate* the tossing of a fair coin, we must replace it with an experiment having two equally likely outcomes. The program, shown below, for your TI-82 graphics calculator simulates the tossing of a fair coin by selecting a random number from the interval, $0 \le x < 2$ and taking its integral part (which will be 0 or 1). Since 0 and 1 are equally likely, we can associate the outcome 0 with tails and the outcome 1 with heads. This program asks for the number of trials (tosses) requested, runs these trials and then reports the number of heads obtained.*

To create program **COINTOSS**, press: [PRGM] [◀] [ENTER] and key in : COINTOSS one letter at a time and press [ENTER].

To access **Input**, press: [PRGM] [▶] [ENTER]

To access **For**, press: [PRGM] [4] [ENTER]

To access **Disp**, press: [PRGM] [▶] [3]

```
PROGRAM:COINTOSS
:Input "HOW MANY
TOSSES?",N
:0→H
:For(J,1,N,1)
:iPart 2rand→X
:If X>0
:H+1→H
:Disp H
:End
:Disp "NUMBER OF
HEADS ",H
:■
```

To access the arrow, press [STO▶]

The **iPart** and the **rand** commands are accessed by pressing [MATH] [▶] and [MATH] [◀] respectively.

To access > press: [2nd] [TEST]

*For more instruction on how to enter and create programs, see our publication: *Programming and Programs for the TI-82 Graphics Calculator.*

WORKED EXAMPLES

WORKED EXAMPLE 1

Enter into your TI-82 calculator, the program, **COINTOSS**.
    a) Run COINTOSS for n tosses of a fair coin when n is equal to: 50, 100, 200, 500 and 1000.
       Create a table to display n and the number of heads for each n.
    b) Calculate the relative frequency of heads in each case and display in your table.
    c) Make a scatterplot of the number of heads vs n.
    d) Determine the experimental probability of "heads" from your data.

Solution

To enter the program, we proceed as described on page 38 and press **2nd** [QUIT] to exit the programming mode.

a) To create a 3-column table with headings, L_1, L_2 and L_3 where L_1 is the number of tosses, L_2 the number of heads and L_3 the relative frequency of heads, press **STAT** **ENTER** . Clear columns labeled L_1, L_2 and L_3. Then in column L_1 enter the list 50, 100, 200, 500 and 1000 as shown in the display.

or whatever the number of
program **COINTOSS** in the list.

To run **COINTOSS** for 50 tosses, press: **PRGM** **1** **ENTER**

In response to the prompt, **HOW MANY TOSSES?** we enter **50** **ENTER** . We obtain the display on the right showing that 27 of the tosses yielded heads. To enter 27 in the list L_2, we press: **STAT** **ENTER** and use the cursor keys to position the cursor in column L_2 to enter 27.

To run COINTOSS for $n = 100$, we press **2nd** [QUIT] **ENTER**
Repeating this procedure for $n = 100, 200, 500$ and 1000, we obtain the table in the display.

b) To display the relative frequencies in column L_3, we quit the STAT mode and press:

2nd [L_2] **÷** **2nd** [L_1] **STO▸** **2nd** [L_3] **ENTER**

To return to the table, we press: **STAT** **ENTER** and obtain the display on the right.

We observe that the relative frequencies seem to approach 0.5 as the number of tosses, n, increases.

c) We set the window variables to $0 \le x \le 1000$; $0 \le y \le 1000$ and press: **2nd** [STAT PLOT]
    We define a plot by selecting the scatterplot option and defining
    the **Xlist** and **Ylist** as L_1 and L_2 respectively.
    Upon pressing **GRAPH** we obtain the display on the right.
    The points in the scatterplot cluster along a straight line of slope 0.5.

d) To obtain the experimental probability of the outcome, HEADS, we plot the relative frequency of HEADS as the number of tosses, n, increases. We obtain the display on the right which shows visually what we saw in the third column of the table; i.e. that the relative frequency of HEADS seems to approach 0.5 as the number of tosses, n, increases.

In all we simulated the toss of a coin 1850 times and observed 946 heads, so the experimental probability of heads obtained in this simulation is 946/1850 or 0.51... Of course, different replications of this simulation will yield different results although most of the time the relative frequency will be close to 0.5.

Number of Tosses
Frequency of Heads
Relative Frequency

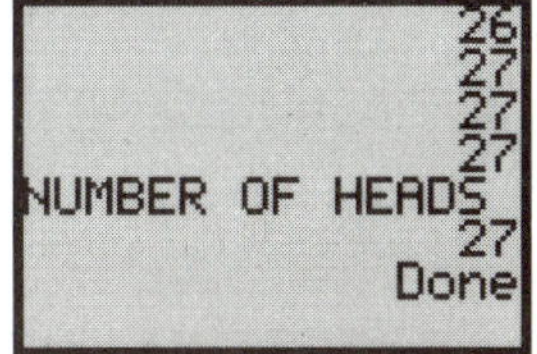

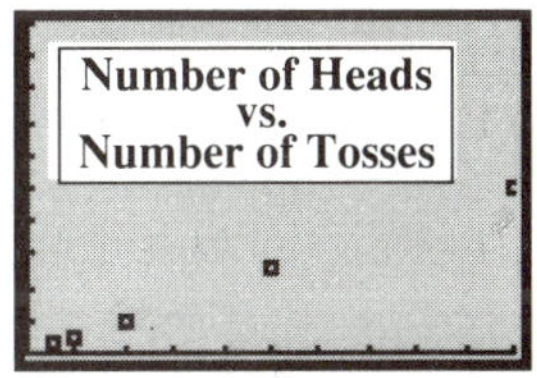
Number of Heads
vs.
Number of Tosses

Number of Heads

Number of Tosses

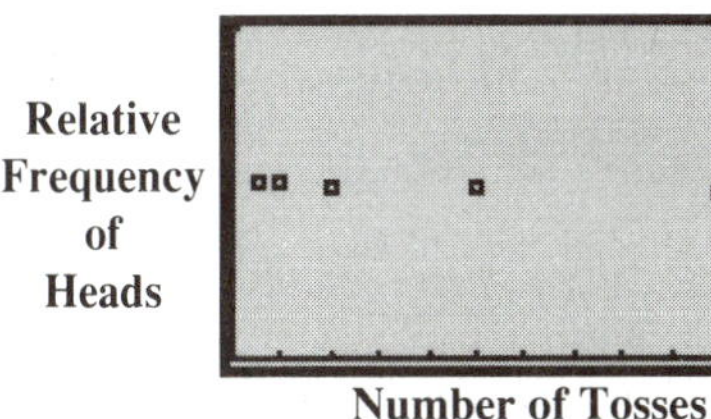
Relative Frequency of Heads

Number of Tosses

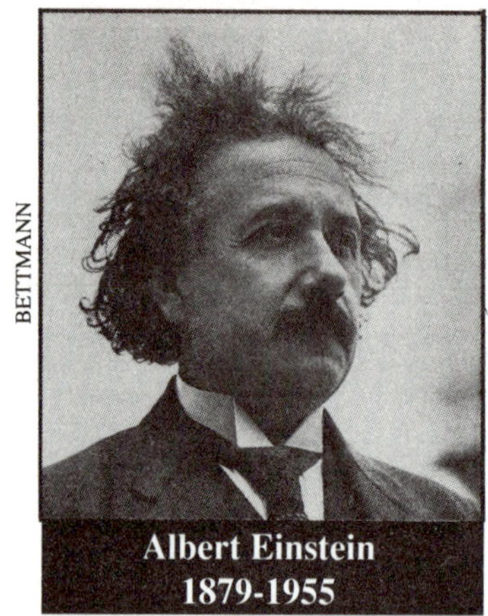

Albert Einstein, the creator of the Special and General Theories of Relativity, revolutionized modern physics. His discovery of the equivalence of mass and energy, in 1905 captured in his famous equation, $E = mc^2$, paved the way for the generation of nuclear energy — a discovery which promises significant benefits to humanity as fossil fuels are depleted. His re-formulation of Newton's laws of motion and his expression of gravitational fields as mathematical tensors, resolved some unanswered questions which had confounded classical physics. Although revolutionary in his approach to physics, Einstein never accepted the interpretations of the new Quantum Physics which called into question strict causality and introduced a dimension of chance into physics. His skepticism is evident in his oft quoted statement,

I shall never believe that God plays dice with the world.

The role of chance in physics and in life itself has fascinated humans since the beginning of history. Dice carved from bones or the horns of animals have been used in games of chance. The six faces of the cube have been considered equally likely outcomes when the cube is rolled and consequently dice are used in most modern board games. When a large number of rolls of a die are required, it is useful to simulate the roll of a die with the random draw of a number from the set, $\{1, 2, 3, 4, 5, 6\}$. The program, DIETOSS, asks for input regarding the number, N, of tosses of a die which you wish to simulate. It draws an integer at random from the interval $[1, 6]$ using the **rand** command on the PRB (probability) submenu of the MATH menu of your TI-82. It then displays a histogram[1] showing the frequency of each outcome on the N "tosses".

WORKED EXAMPLE 2

Enter into your TI-82 calculator, the program, **DIETOSS**.
   a) Run DIETOSS to simulate 100 tosses of a die.
   b) Trace along the histogram to determine the frequency of the outcome, which occurs most often.
   c) Display a table showing the frequency of each outcome.

Solution

To create program **DIETOSS**, we press: [PRGM] [◄] [ENTER]

and key in: DIETOSS one letter at a time and press [ENTER].

To access **dim**, press: [2nd] [LIST] [3]

To access **For**, press: [PRGM] [4] [ENTER]

To access **If**, press: [PRGM] [ENTER]

To access **End**, press: [PRGM] [7] [ENTER]

a) We run DIETOSS and enter 100 in response to the prompt. In the case shown here, we obtain the histogram below left.

b) We press: [TRACE] and move the cursor to the tallest bar. The display shows that the frequency of outcome 5 was 25.

c) When we press: [STAT] [ENTER] we get the table below.

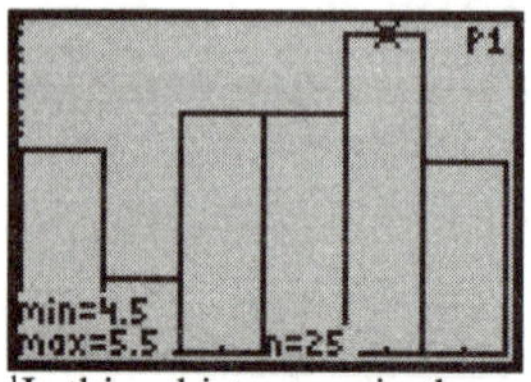

```
PROGRAM:DIETOSS
:Input "HOW MANY
TOSSES?",N
:0→Xmin:6→Xmax
:0→Ymin
:Fix 0
:{1,2,3,4,5,6}→L
1
:6→dim L₂
:0L₂→L₂
:For(I,1,N,1)
:iPart 6rand+1→X
:For(J,1,6,1)
:If X=J
:L₂(J)+1→L₂(J)
:End
:End
:max(L₂)+1→Ymax
:Plot1(Histogram
,L₁,L₂)
:FnOff
:PlotsOn 1
:ZoomStat
```

[ZOOM] [9] [ENTER]

This sets the mode to 0 decimal digits. To access this press: [MODE]

Press: [STO►] [2nd] [L₁]

The **iPart** and the **rand** commands are accessed by pressing [MATH] [►] and [MATH] [◄] respectively.

[2nd] [STAT PLOT] [ENTER]

[2nd] [STAT PLOT] [►]

[4] [ENTER]

[2nd] [STAT PLOT] [5] [1] [ENTER]

L1	L2	L3
1	16	------
2	6	
3	19	
4	19	
5	25	
6	15	

L₂(₅)=25

[1]Is this a histogram in the pure sense? Who cares? *Rigor*mortis may set in if we worry about it.

1. Explain the difference between experimental probability and theoretical probability.

2. When a paper cup is tossed, there are 3 possible outcomes. The cup can land…

a) upright   b) upside down   c) on its side

  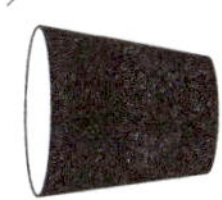

Explain how you would obtain the experimental probability of each outcome.

3. The probability that a randomly chosen American citizen will ultimately die from heart disease is about 0.43.
   a) How is this probability determined?
   b) Is this a theoretical probability or an experimental probability? Explain why this probability is used.

Enter the program, COINTOSS, given on page 38 by following the procedures given in worked example 1. Use this program to help you complete exercises 4 and 5.

4. a) Run your program to simulate 100 tosses of a fair coin. Record the frequency of heads. Repeat this experiment 9 more times and record the frequency of heads in each case.

   b) Find the median frequency of heads for the 10 simulations. Explain why one might expect to find a median frequency close to 50.

5. a) Run your program to simulate 50, 100, 200, 500 and 1000 tosses of a fair coin. Display the number of tosses, frequencies and relative frequencies in a table as in worked example 1.

   b) Compare the relative frequencies of heads in your table with those in worked example 1 for 50 tosses and for 1000 tosses. Are the relative frequencies closer for $n = 50$ or for $n = 1000$? Is this what you would expect? Explain.

6. a) The program below simulates n trials of the game described in the *Problem of the Points* on pages 32 and 33. Enter this program into your calculator and run it for 100 games. Record the number of times that A wins.

b) Repeat the procedure in part "a" 9 more times and record the number of times that A wins each time.

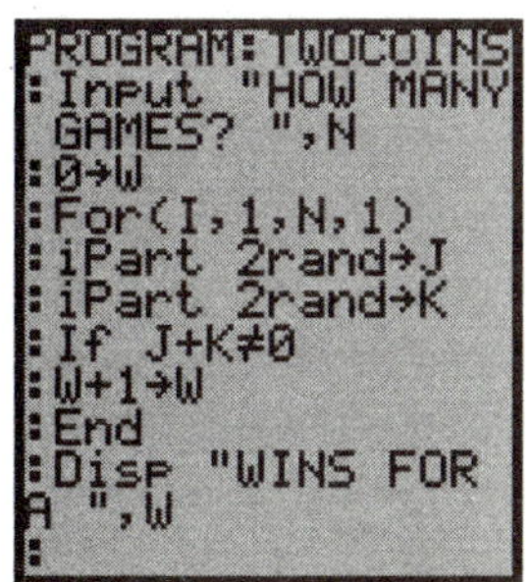

c) Which model, the one developed by Roberval, or the one developed by Pascal yields a theoretical probability which is closer to the experimental probability that A wins?

7. a) If H and T denote respectively the frequencies of heads and tails in the toss of a fair coin, estimate the value of $|H - T|$ after 100, 500 and 1000 tosses of a fair coin.
   b) Use your COINTOSS program to determine whether your estimate is reasonable.
   c) Does $|H - T|$ approach 0 as the number of tosses becomes very large? Explain your answer.

8. Enter the program DIETOSS into your calculator. (See worked example 2.)
a) Run DIETOSS to simulate 100 tosses of a die. Trace along the histogram to determine the frequency of each outcome.

b) Display the frequency and the relative frequency of each outcome in a table.

c) Compare the relative frequencies you obtained with those presented in worked example 2. Compare the experimental probability of each outcome with its theoretical probability.

d) Repeat the simulation for 1000 tosses and display in a table, the relative frequencies of all the outcomes. What do you discover about the relative frequencies of the outcomes as the number of tosses increases? (Retain COINTOSS, TWOCOINS and DIETOSS in for use in Exploration 11.)

9. Calculate the theoretical probability of each event on the roll of a die.
   a) an even number occurs
   b) a number greater than 4 occurs
   c) a 5 does not occur
   d) either a 3 or a 6 occurs

Run DIETOSS to simulate 1000 rolls of a die to determine the experimental probability of each of the above events. Compare the theoretical probability and the experimental probability of each event. Compare the theoretical and experimental probabilities for a simulation of 5000 rolls of a die.

Use the COINTOSS simulation to estimate the probability that a family of 7 children has exactly 4 boys. (Assume that both genders are equally likely at birth.)

Compare the results of your simulation with your answer to exercise 8 on page 37.

In 1827 a botanist, Robert Brown, observed through his microscope the erratic motions of pollen dust in water. He hypothesized that the dust particles were under bombardment from the water molecules. This effect has later become known as *Brownian motion*. Since the molecular theory of matter was considered highly speculative, Brownian motion received little attention until the turn of the century.

In 1905, Albert Einstein published four epoch making papers in *Annalen der Physik*. One of those papers presented a mathematical model for Brownian motion. The subsequent extensions of Einstein's work by other mathematicians such as Norbert Wiener produced mathematical models which provided accurate representations of diffusion processes. Today diffusion processes are modeled by simulations such as the *random walk*.

In 1921, Albert Einstein won the Nobel Prize, not for the mass-energy equivalence discovery, nor for his discovery of the Theory of Relativity, nor for his work on Brownian motion. For what work did Einstein receive his Nobel prize?[1] (See footnote.)

In the one-dimensional random walk, a particle bombarded by a molecule is moved to either the right or the left with equal probability. The question to be investigated is, *What is the long term behavior of the particle when the number of bombardments becomes very large?* Will the particle end up close to where it began, will it eventually move away from its initial point never to return, or will it oscillate back and forth between these extremes? To simulate this problem, we consider a particle which is situated at 0 on a number line. We then toss a coin and move 1 unit right if the coin shows heads and 1 unit left if the coin shows tails. Where do you think the particle will be after 100 tosses of the coin? Where do you think it will be after 1000 tosses?

Where Will the Particle Be After 5000 Tosses of the Coin?

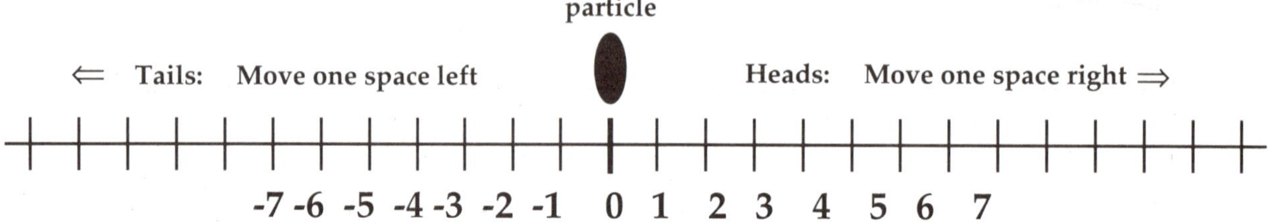

Run COINTOSS to simulate 100, 1000 and 5000 bombardments of the particle.

Display the final positions of the particle in a table.

Describe how the probability that the particle will finish near 0 changes as the number of tosses becomes vey large.

Solution

When we run COINTOSS with N = 100, 1000 and 5000 respectively, we obtain the frequencies of heads shown in the table in column L_2. The corresponding final positions of the particle are shown in column L_3 where $L_3 = 2L_2 - L_1$.

The numbers in column L_3 suggest that the final distance from 0 increases with N. As N increases, there is an increased number of final positions which the particle can occupy, so the probability that it will occupy a final position close to 0 decreases.

L1	L2	L3
100	52	4
1000	507	14
5000	2471	-58
------	------	

L1(4)=

---

[1] Einstein's 1905 paper which gave a mathematical model explaining the photoelectric effect won him the Nobel Prize in 1921.

WORKED EXAMPLES

WORKED EXAMPLE 2

Run the simulation, COINTOSS to estimate the probability that there will be exactly 3 girls in a family of 5 children.
Compare this estimate with the theoretical probability calculated in worked example 2, page 36.

Solution

Since both genders are equally likely at birth, we can simulate the selection of a gender by the toss of a coin. Getting exactly 3 girls out of 5 children has the same probability as achieving 3 heads out of 5 tosses of a fair coin. Therefore, to simulate the assignment of genders to 5 children, we run COINTOSS for 5 tosses. To do this, press: **PRGM** **1** **ENTER**

or whatever the number of program **COINTOSS** *in the program list.*

In response to the prompt, HOW MANY TOSSES?, press: **5** **ENTER**
Almost immediately the calculator displays: NUMBER OF HEADS
followed by a number between 0 and 5 on the line below.
This represents the number of girls in that selection of 5 children. We run this experiment 99 more times, by pressing **ENTER** **5** **ENTER** each time and forming a tally like the one on the right.

Number of Trials	Freq of outcome 3	Relative Frequency
ᛏᛏᛏ ᛏᛏᛏ ᛏᛏᛏ ᛏᛏᛏ ᛏᛏᛏ ᛏᛏᛏ ᛏᛏᛏ ᛏᛏᛏ ᛏᛏᛏ ᛏᛏᛏ ᛏᛏᛏ ᛏᛏᛏ ᛏᛏᛏ ᛏᛏᛏ ᛏᛏᛏ ᛏᛏᛏ ᛏᛏᛏ ᛏᛏᛏ ᛏᛏᛏ ᛏᛏᛏ	ᛏᛏᛏ ᛏᛏᛏ ᛏᛏᛏ ᛏᛏᛏ ᛏᛏᛏ ᛏᛏᛏ ᛏᛏᛏ ///	38 out of 100 = 0.38

The simulation gives an experimental probability of about 0.38. The theoretical probability calculated in worked example 2, page 36 is 0.3125. By increasing the number of trials, we would probably approximate the theoretical probability even more closely.

The program, BINOMIAL, in worked example 3, automates the tedious tally process in worked example 2. BINOMIAL displays in a histogram the frequency of each outcome in the tossing of N coins repeated M times.

WORKED EXAMPLE 3

Enter program, BINOMIAL, into your calculator. Run BINOMIAL to estimate the probability that there will be exactly 3 girls in a family of 5 children. Compare your estimate with the estimate in worked example 2 and with the theoretical probability.

Solution

We enter program, BINOMIAL, using the menus described in the worked examples of this and the previous exploration. When we run BINOMIAL, we enter 5 in response to the prompt, HOW MANY COINS?. We enter 100 in response to HOW MANY TOSSES?.

Within 30 seconds, the display shows the histogram below. There are 6 bars representing the frequencies of the 6 possible outcomes: 0 heads, 1 head, 2 heads,…5 heads. (The first bar is obscured by the text on the display.) We press **TRACE** and move the cursor to the fourth bar which represents the frequency of exactly 3 heads. The display shows $n = 29$, so the experimental probability is 0.29 compared to 0.3125, theoretical probability and 0.38 in worked example 2. Press **STAT** **ENTER** to display the table of outcomes (L_1), frequencies (L_2) and relative frequencies (L_3).

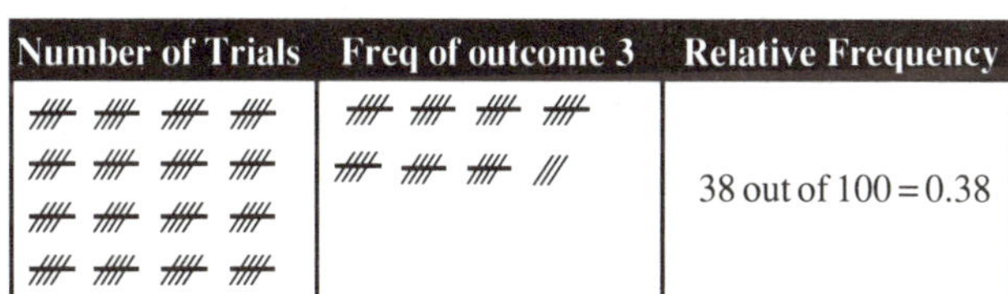

```
PROGRAM:BINOMIAL
:Float
:Input "HOW MANY
 COINS? ",N
:Input "HOW MANY
 TOSSES?",M
:-1→Xmin:N+1→Xma
x
:N+1→dim L₁
:N+1→dim L₂
:seq(X,X,0,N,1)→
L₁
:0L₂→L₂
:For(K,1,M,1)
:0→H
:For(J,1,N,1)
:iPart 2rand→Y
:If Y>0
:H+1→H
:End
:For(L,0,N,1)
:If H=L
:L₂(L+1)+1→L₂(L+
1)
:End
:End
:max(L₂)+1→Ymax
:L₂/M→L₃
:Plot1(Histogram
,L₁,L₂)
:FnOff
:PlotsOn 1
:ZoomStat
```

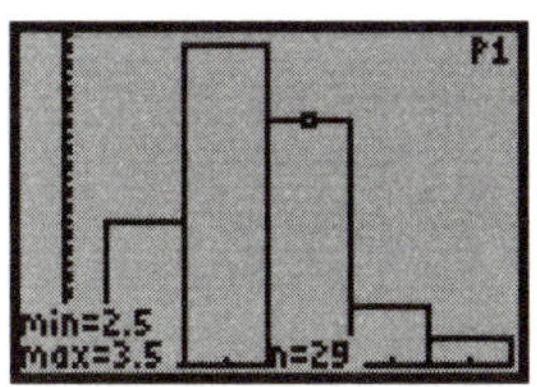

L₁	L₂	L₃
0	6	.06
1	17	.17
2	38	.38
3	29	.29
4	7	.07
5	3	.03

L₁(7)=

The previous examples show how we can use a coin toss to simulate processes involving an aggregate of successive events when each event has two equally likely outcomes. The next example shows how we can use a simulation to estimate the area of an ellipse. To perform this simulation, we use program, **ELLIPSE**, which selects points randomly from the rectangle defined by: $-A \le x \le A$; $-B \le y \le B$ and records the proportion of those points which lie inside or on the ellipse with equation: $B^2x^2 + A^2y^2 = A^2B^2$

WORKED EXAMPLE 4

Run program ELLIPSE to estimate the area of the ellipses with these equations:

a) $9x^2 + 16y^2 = 144$ b) $x^2 + 4y^2 = 4$ c) $9x^2 + y^2 = 9$

Use your answers to conjecture a formula for the area of an ellipse with equation: $B^2x^2 + A^2y^2 = A^2B^2$

Solution

a) When we run **ELLIPSE**, we enter 4 and 3 respectively in response to the prompts **A?** and **B?**, because these are the values of the x and y-intercepts. We enter 500 in response to the prompt, **HOW MANY POINTS?**. This causes **ELLIPSE** to select 500 random points from the rectangle, $-A \le x \le A$; $-B \le y \le B$ and test each one to determine whether it is inside the ellipse. As the points are selected, they are displayed on the screen as shown in the display below.

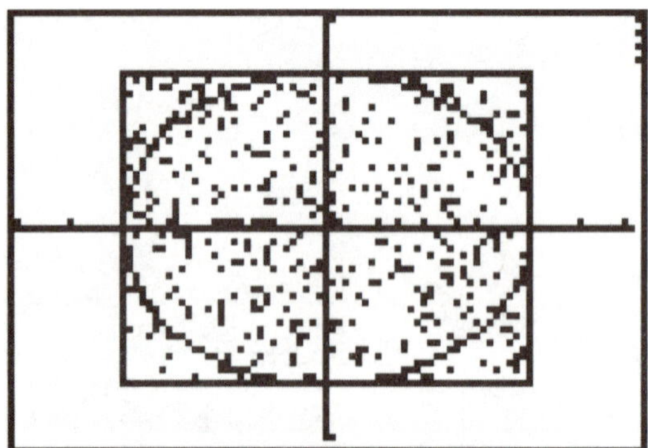

When the cursor in the upper right corner changes from the busy to the pause mode, we press **ENTER** and we obtain the display shown below.

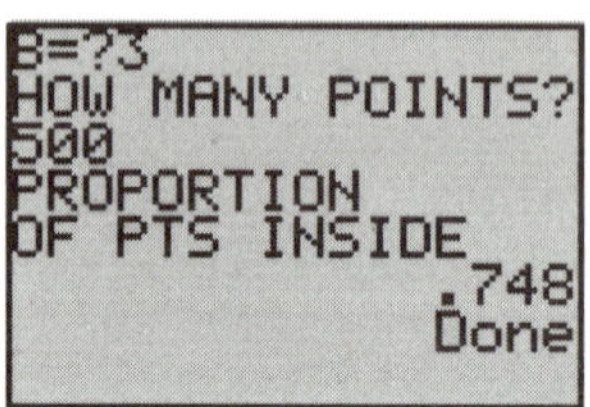


```
PROGRAM:ELLIPSE
:prgmDEFAULTS
:Prompt A,B
: -A-1→Xmin
:A+1→Xmax
: -B-1→Ymin
:B+1→Ymax
:Input "HOW MANY
POINTS? ",N
:0→S
:"(B/A)√(A²-X²)"
→Y1
:"(-B/A)√(A²-X²)
"→Y2
:DispGraph
:Line(-A,-B,A,-B
)
:Line(-A,B,A,B)
:Line(-A,-B,-A,B
)
:Line(A,-B,A,B)
:For(I,1,N,1)
:2Arand-A→X
:2Brand-B→Y
:Pt-On(X,Y)
:If B²X²+A²Y²<A²
B²
:S+1→S
:End
:S/N→T
:Pause
:Disp "PROPORTIO
N"
:Disp "OF PTS IN
SIDE",T
:
```

This display indicates that of all the points randomly selected, 74.8% of them were inside the ellipse. Since all points are equally likely, it would seem that the area of the ellipse must be approximately 74.8% of the area of the bounding rectangle. The area of the bounding rectangle is 4**AB** or 4(4)(3) = 48 square units, so the area of the ellipse is about 0.748 × 48 or 35.9 square units. (The actual area of the ellipse is 37.7 square units.)

b) We repeat the procedure in part a) with **A** = 2; **B** = 1 and **N** = 300. We obtain the displays from which we obtain an estimated area of 0.812 × 8 or 6.5 square units.

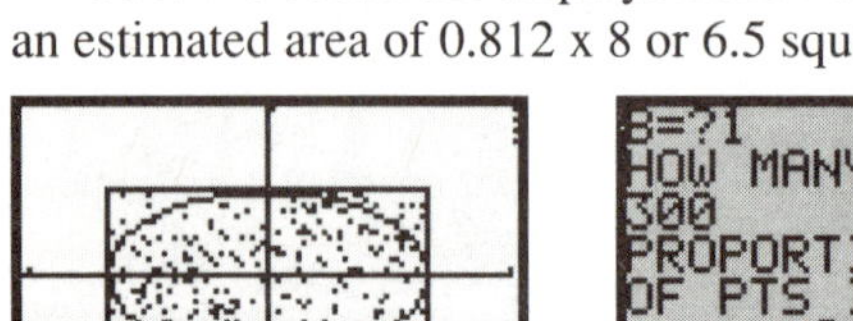
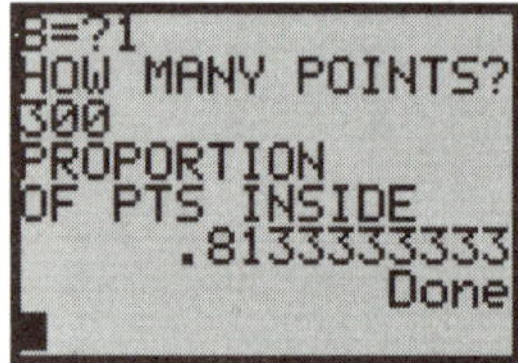

b) We repeat the procedure in part a) with **A** = 1; **B** = 3 and **N** = 300. We obtain the displays from which we obtain an estimated area of 0.817 × 12 or 9.8 square units.

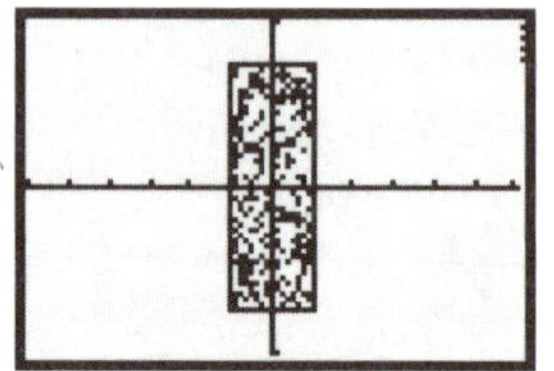
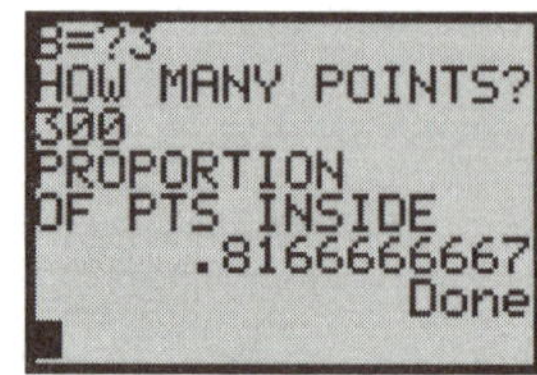

When we run **ELLIPSE** for large numbers of points, we find that the proportion approaches 78.5%. The estimate of the area of the ellipse is therefore $0.785(4AB) \approx 3.14AB$. Look familiar?

1. What is a simulation? Explain how simulations are used to obtain experimental probabilities.

2. A simulation is based on the idea that the experimental probability of each outcome will be approximately equal to its theoretical probability (if one exists). However, the laws of chance dictate that occasionally there will be a significant gap between an experimental probability and the corresponding theoretical probability. Is there any way to be sure that such a discrepancy does not occur? Explain.

Use **COINTOSS, DIETOSS** or the **BINOMIAL** simulation to help you complete exercises 3 through 8.*

3. a) Estimate the probability of guessing exactly 7 correct answers on a 10-item true-false test.
b) Estimate the probability of guessing at least 7 correct answers on a 10-item true-false test.

4. a) Estimate the probability that there are exactly 4 girls in a family of 6 children.
b) Estimate the probability that there are at least 4 girls in a family of 6 children.

5. Each item on a 20-item multiple choice test offers 6 possible answers. Estimate the probability that a student will guess correctly on more than one quarter of the items.

6. In a particular occupation, one third of all employees are female. The Acme Company has 82 employees in that occupation but only 21 of them are female. Estimate the probability that this gender balance could have occurred by chance.

7. An airline books 96 passengers on its jet liner which has capacity for only 85 people. If there is only one chance in 6 that a passenger will not show for the flight, what is the probability that more than 85 of the 96 passengers will show?

8. The members of the Old Timers Hockey League draw slips of paper before each game to determine which of the 6 positions each will play. Estimate the probability that a particular member will have played all 6 positions after 20 games.

9. a) Run **ELLIPSE** to estimate the area of the ellipses defined by these equations.

i) $7x^2+2y^2=14$ ii) $3x^2+9y^2=27$ iii) $12x^2+7y^2=97$

b) Compare you estimates with the actual area given by:

Area$=\pi AB,$ where A and B are the x and y-intercepts

Explain how your estimate changes as the number of points selected is increased.

10. a) Run **ELLIPSE** with A = √2 , B = √7 and N = 1000. Use the output to estimate the area of the ellipse defined by the equation: $7x^2+2y^2=14$.

b) Modify **ELLIPSE** to create program **HCYCLOID**, which estimates the area inside the hypocycloid defined by the equation:
$$x^{2/3}+y^{2/3}=c \quad \text{where } c \text{ is a constant}$$

c) Run **HCYCLOID** to estimate the area under the hypocycloid with $c = 3$.

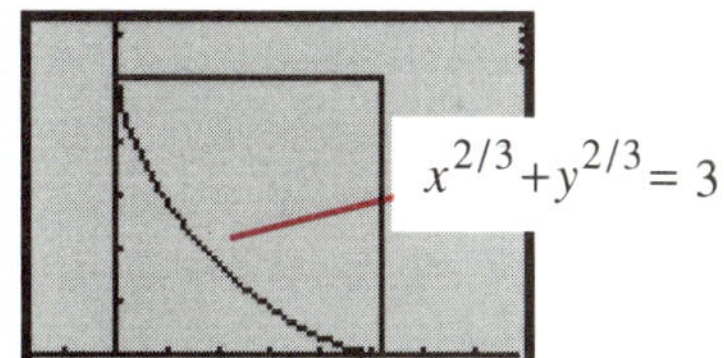

d) The formula for the area under a hypocycloid is:
Area $= k\pi c^3$ where c is the constant in the equation Use **HCYCLOID** to estimate k.

11. a) Run **BINOMIAL** for N = 10000 to estimate the probability that there will be exactly 4 boys in a family of 7 children. (This will take a couple of hours to run, so do it before you go to bed and let the program run while you sleep. You can harvest the results in the morning.)

b) Compare your experimental probability with the theoretical probability you calculated in your answer to exercise 8 page 37.

12. a) Modify **DIETOSS** to create **MTOSS**, a program which creates the same kind of histogram as **DIETOSS**, but for an M-sided die.

b) An employer discovers that 496 of the 2017 absences reported last month, occurred on Fridays. Run **MTOSS** to estimate the probability that this would occur from illness alone, assuming that illness occurs on any of the five work days of the week with equal probability.

CHALLENGE

In the *Two Dimensional Random Walk*, a particle is bombarded by molecules. The particle moves one space up, down, right or left, each with probabiliy of 1/4. Use **MTOSS** (see exercise 12) to estimate how far the particle is likely to be from its initial position after 5000 bombardments.

*For instruction on how to enter and create programs, see our publication: *Programming & Programs for the TI-82 Graphics Calculator*

HOW WOULD YOU COMPUTE THE PROBABILITY OF WINNING AT SOLITAIRE?

Stanislaw Ulam, one of the co-developers of the atomic bomb, was stationed in Los Alamos, New Mexico, during the 1940's. Between bomb tests, there was sufficient time for Ulam to engage himself in the game of solitaire. During one such involvement, he pondered the question,

In what fraction of all the games of solitaire will a player complete to the last card?.

That is, what is the probability of winning at solitaire? Since this problem is too complex (i.e. too many cases) to be solved using theoretical probability computations, Ulam programmed one of the early mainframe computers to simulate solitaire and he ran hundreds of trials. He then computed the proportion of times that the computer won. In this way he found the relative frequency of the "win" outcome and estimated its probability. Such a procedure for estimating the probability of an outcome by computing the experimental probability of that outcome in a simulation is called a *Monte Carlo technique*. The simulations in Exploration 12 which were used to calculate experimental probabilities are examples of Monte Carlo techniques. Such techniques have been in use prior to this century but it was Ulam, von Neumann and others who formalized these methods and established them as a legitimate branch of mathematics.

The computer technology which has emerged in the last few decades has increased the power and scope of these methods to a hitherto unimagined extent. Some of the simple programs shown here for the TI-82 suggest new approaches to problems which were previously intractable.

WORKED EXAMPLE 1

Modify **DIETOSS** to create **MTOSS**, a program which creates the same kind of histogram as **DIETOSS**, (See page 40) but for an M-sided die with all sides equally probable outcomes.

An employer discovers that 496 of the 2017 absences reported last month, occurred on Fridays. Run **MTOSS** to estimate the probability that this would occur from illness alone, assuming that illness occurs on any of the five work days of the week with equal probability.

Solution

Program **MTOSS** is shown on the right. We merely replace the number 6 (for 6 sides or faces) with the letter **M** (for **M** sides or faces) in program **DIETOSS**. We also change the definition of L_1 to give it dimension **M**.

Since sickness is equally likely to occur on any one of the 5 days, Monday through Friday, we can simulate the selection of a sick day by the toss of a 5-sided die with faces marked M, T, W, Th and F. In our simulation, we toss the die 2017 times and observe whether outcome F occurs as often as 496 times. We repeat this experiment 20 times and record the fraction of times outcome F occurred 496 or more times. This gives us an estimate of the probability that this number of absences on Friday could occur by chance.

```
PROGRAM:MTOSS
:Input "HOW MANY
 TOSSES?",N
:Input "HOW MANY
 SIDES? ",M
:0→Xmin:M+1→Xmax
:0→Ymin
:Fix 0
:seq(Z,Z,1,M,1)→
 L₁
:M→dim L₂
:0L₂→L₂
:For(I,1,N,1)
:iPart Mrand+1→X
:For(J,1,M,1)
:If X=J
:L₂(J)+1→L₂(J)
:End
:End
:max(L₂)+1→Ymax
:Plot1(Histogram
 L₁,L₂)
:FnOff
:PlotsOn 1
:ZoomStat
:
```


WORKED EXAMPLE 1

Solution (cont'd)

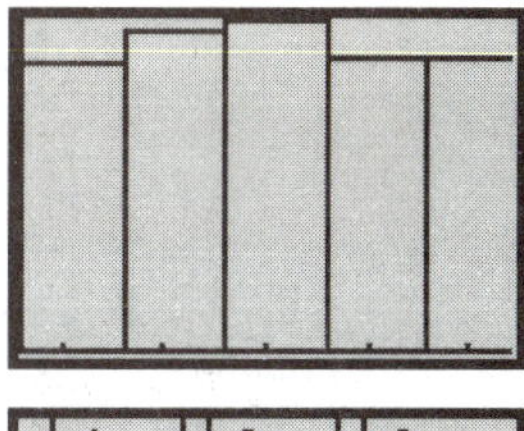

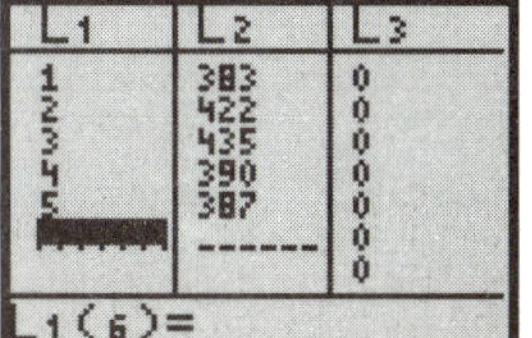

When we run **MTOSS**, with N = 2017 (number of trials) and M = 5 (number of possible equally likely outcomes on each trial), we obtain a histogram like the one shown in the display. The frequency of outcome, "Friday" is represented by the rightmost bar on the histogram. When we press, **STAT** **ENTER** we see from the table that on this replication of the simulation, only 387 sickdays fell on Friday. That is, on the first trial, the number of Friday sickdays was less than 496. To estimate the fraction of times the number of Friday sickdays will be less than 496, we run **MTOSS** 20 times. The number of Friday sickdays on replications 2 through 20 were: 389, 421, 395, 365, 370, 427, 385, 409, 406, 366, 389, 405, 430, 417, 401, 372, 411, 433 and 414. None of these is as large as 496.

It seems that that there is a small probability that 496 sickdays would fall on Friday on the basis of chance alone. However, before introducing major policies to deal with these absences, management should collect data for several months to determine whether this pattern is maintained and to investigate other possible reasons for disproportionate absence on Fridays.

WORKED EXAMPLE 2

Each box of Brainfood Cereal contains a porcelain figurine of one of world's 10 most celebrated scientists. The 10 figurines which constitute a complete set are randomly distributed in equal quantities. What is the average (mean) number of boxes of cereal which a person could expect to buy in order to obtain a complete set of figurines?

Solution

How **FIGURINE** simulates the purchase of cereal boxes.

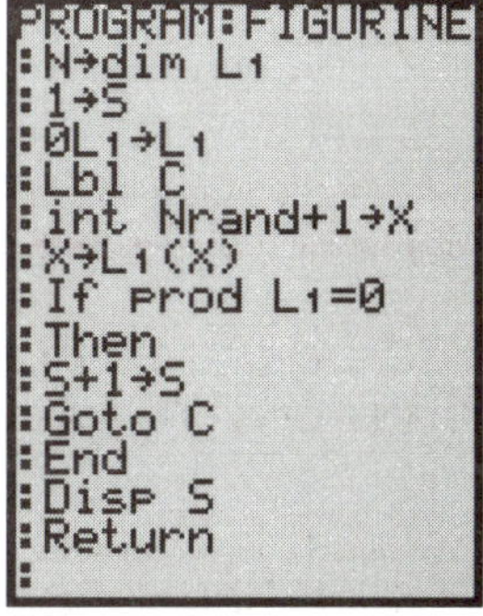

Program **FIGURINE** simulates the random selection of integers from the set of N integers, $\{1, 2, 3, \ldots N\}$, where N is the number of "figurines" in a complete set. When integer, X is selected by the program, it is stored to $L_1(X)$, the Xth term of list L_1. We have a complete set when each integer between 1 and N has been selected at least once; that is, all terms of L_1 are non-zero. Until this occurs, the product of the terms of L_1 is zero, and the random drawing continues. The variable, S, counts the number of items selected. When the set is complete; i.e. **prod L_1 > 0**, the display shows, S, the number of selections required to get a full set.

How **FIGURINE** is used

Before running **FIGURINE**, we store 10 to the variable N, to indicate that there are 10 figurines in a complete set. When we ran **FIGURINE** once, we obtained the display shown here. That is, on one replication of the simulation, it took 32 cereal boxes before we had the complete set of 10 figurines. Is this typical, or was this just an unlucky draw?

To determine whether 32 is close to the average (mean) number of boxes we would have to purchase, we could run **FIGURINE** 50 times and take the mean number of boxes required. Alternatively, we could write a program such as **FULLSET**, (shown here) to call **FIGURINE** as a subroutine and replicate it T times.

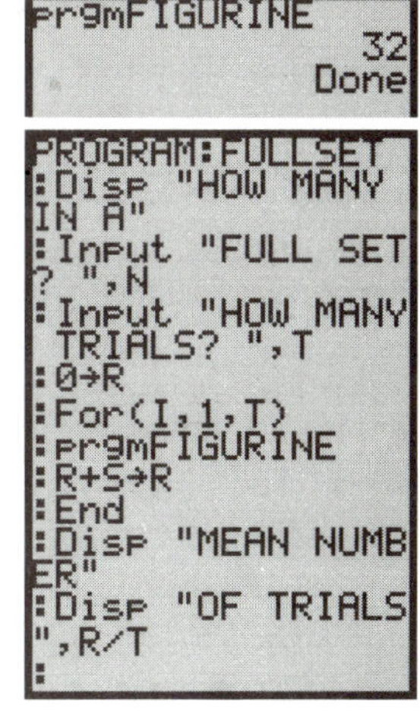

When we run **FULLSET**, we enter 10 and 20 in response to the prompts for the number of a full set and the number of trials. The display shows the output for a single run. We conclude that on average, we would have to purchase about 30 boxes of cereal to get the full set of 10 figurines.

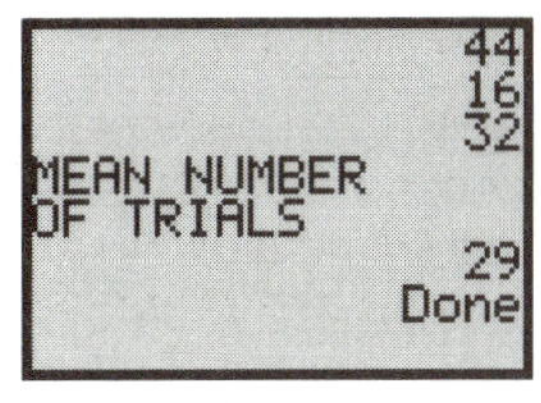

WORKED EXAMPLE 3

What is the probability that two people in a group of N people have coincident birthdays; i.e. birthdays on the same day of the same month (but not necessarily the same year) ?

Run program, **COINCIDE**, (with subroutine, **BIRTH**) to estimate this probability when $N = 24$.

Solution

Suppose we have a group of N people. We ask each person to write his[1] birthdate on a slip of paper which we place in a hat. As we draw these slips of paper from the hat, we assume each of the 366 dates is equally likely (although Feb 29 is less likely than the rest).

We shall calculate first the probability that there are no coincident birthdays.

On the first draw there can be no coincidence. (It's the first one considered.)

On the second draw, the probability that the date chosen is different from the first is: $\dfrac{365}{366}$

On the third draw, the probability that the date differs from the first two is: $\left(\dfrac{365}{366}\right)\left(\dfrac{364}{366}\right)$

-
-
-

On the Nth draw, the probability that the date differs from the previous (N-1) is:

$$\left(\frac{365}{366}\right)\left(\frac{364}{366}\right)\left(\frac{363}{366}\right)\ldots\ldots\ldots\ldots\left(\frac{366-(N-1)}{366}\right)$$

Therefore the probability of coincidence is: $1-\left(\dfrac{365}{366}\right)\left(\dfrac{364}{366}\right)\left(\dfrac{363}{366}\right)\ldots\ldots\ldots\ldots\left(\dfrac{366-(N-1)}{366}\right)$

To calculate the probability when N = 24, we substitute into the formula to obtain:

Probability of coincidence = $1-\left(\dfrac{365}{366}\right)\left(\dfrac{364}{366}\right)\left(\dfrac{363}{366}\right)\ldots\ldots\ldots\ldots\left(\dfrac{344}{366}\right)\left(\dfrac{343}{366}\right)$

To evaluate this expression on the TI-82, we use the **prod** and **seq** functions from the LIST menus and enter the expression as in the display below left. We observe that the probability of coincidence with 24 people ≈ 0.57—that is, coincidence is more likely than not!

```
PROGRAM:BIRTH
:0[A]→[A]
:1→I
:{61,6}→dim [A]
:Lbl A
:int 61rand+1→X
:int 6rand+1→Y
:If [A](X,Y)=0
:Then
:1→[A](X,Y)
:IS>(I,N)
:Goto A
:Disp "NO COINCI
DENCE"
:Else
:S+1→S
:Disp "COINCIDEN
CE"
:End
:Return
```

```
PROGRAM:COINCIDE
:Input "HOW MANY
TRIALS?",T
:Input "HOW MANY
PEOPLE? ",N
:Float
:0→S
:For(J,1,T)
:prgmBIRTH
:End
:Disp "FRACTION
OF"
:Disp "TRIALS WI
TH"
:Disp "COOINCIDE
NCE ",S/T
```

To simulate the birthday problem, we entered both programs, **BIRTH** and **COINCIDE** and ran **COINCIDE** with N = 24 and "**NUMBER OF TRIALS**" = 200. We obtained the display on the right indicating a probability of coincidence of 0.53, which is close to the theoretical probability, 0.57.

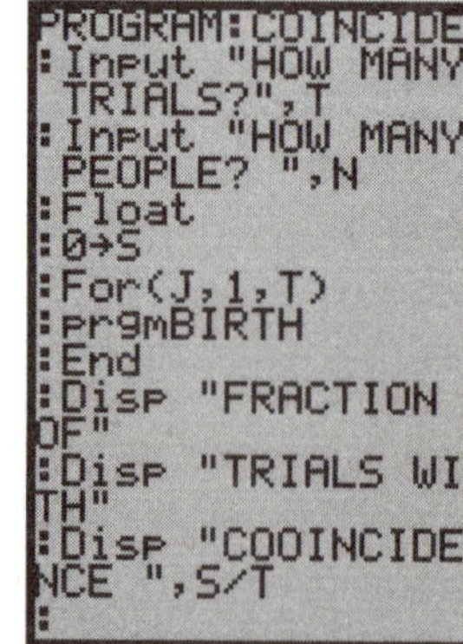

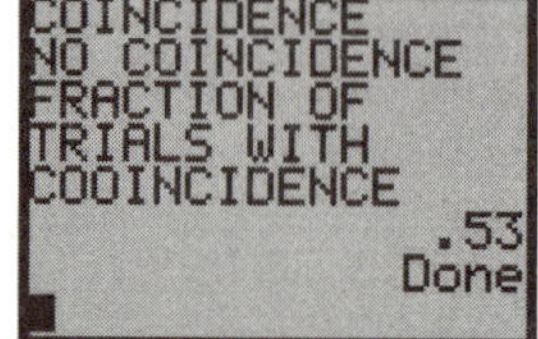

---

[1] This is the generic "his" and carries no gender connotation.

1. Describe what is meant by a Monte Carlo technique.

2. Each item on a 25-item multiple choice test offers 4 possible answers. Run MTOSS to estimate the probability that a student would score at least 10 correct if the student guesses randomly the answer to each item.

3. The probability that an archer will shoot into the inner circle of a target is 0.2. Estimate the probability that the archer will have at least 4 arrows out of 20 land within the inner circle.

4. At a crap table in a famous casino, the dice showed snake eyes (that is; both dice showed a single dot), 30 times on 120 rolls. Estimate the probability of this happening if all 36 outcomes on the roll of the two dice are equally probable?

5. In worked example 3, we discovered that the probability of *no* coincident birthdays in a group of 24 people is given by:

$$\left(\frac{365}{366}\right)\left(\frac{364}{366}\right)\left(\frac{363}{366}\right)\cdots\cdots\cdots\left(\frac{344}{366}\right)\left(\frac{343}{366}\right)$$

Write this expression using factorial notation.
Can you evaluate the expression using the PRB submenu of the **MATH** menu? Explain why or why not.

6. a) Run program **MTOSS** to simulate the probability that at least two people in a group of 5 people will have birthdays in the same month.

b) Calculate the theoretical probability that at least two people in a group of 5 people will have birthdays in the same month. Compare with the experimental probability in part a).

7. Modify program **BIRTH** to estimate the probability that at least two people in a group of N people will have birthdays in the same month. Call your modified program as a subroutine of **COINCIDE**, to estimate this probability using 50 trials.

8. Professor Polyhedron leaves for the University at a random time between 7:00 A.M. and 7:30 A.M. The Daily Planet newspaper arrives at her home at a random time between 6:40 A.M. and 7:10 A.M. Estimate the probability that Professor Polyhedron receives her paper before she leaves the house on at least one day out of every 5.

9. Enter programs, **FIGURINE** and **FULLSET** into your calculator. Run **FULLSET** to estimate the average (mean) number of packages of gum which a baseball fan might expect to purchase to collect the full set of 30 baseball cards. Assume that there is one baseball card in each package of gum and that all baseball cards are equally likely.

10. a) Compute the probability of coincident birthdays for a group of N people where N takes each of these values:
(i) 10   (ii) 15   (iii) 20   (iv) 25   (v) 30   (vi) 35

b) Run **COINCIDE** to determine whether your answers in part a) are reasonable.

c) Create a table with L_1 taking the values 10 through 35 in increments of 5 and L_2 taking the corresponding probabilities of coincident for these values of N.

d) Construct a line graph of L_2 vs. L_1.

11. a) Write a program, **TWODICE**, which simulates N tosses of a pair of dice. On each toss of the dice, **TWODICE** computes the total showing on the two dice and at the end of N tosses, displays a histogram showing the frequency of each total between 2 and 12.

b) Run **TWODICE** for 100 tosses of the dice and record the experimental probability of each sum.

c) Compute the theoretical probability of each sum and compare with the experimental probability in part b).

12. a) We can run program **BINOMIAL** (See page 43) to simulate the tossing of 10 coins. On each toss, there are 11 possible outcomes; 0 heads, 1 head, 2 heads,…10 heads. Run **BINOMIAL** for 500 tosses of 10 coins.

b) Graph the function defined by $y = \dfrac{315}{\sqrt{2\pi}} e^{-\frac{(x-5.5)^2}{5}}$

so that it overlays your histogram. What do you discover? Repeat this procedure for the tossing of 20 coins. Describe how the histogram changes as the number of coins increases.

The program, **BINOMIAL**, simulates M replications of the toss of N coins. Create a program, **DINOMIAL** which simulates T replications of the toss of N dice which are M-sided. (That is, each die has M faces, all of which are equally likely outcomes.) **DINOMIAL** should call MTOSS as a subroutine.

Run **DINOMIAL** to estimate the probability that at least two people in a group of 5 people will have birthdays in the same month. Compare this estimate with your answers to exercises 6 and 7.

To calculate the theoretical probability that exactly 4 of 10 children will be female, we observe that this probability is the same as the probability of achieving exactly 4 heads on the toss of 10 coins.

Using a tree diagram, or otherwise, we reason that the probability of any particular gender configuration such as BBGGBGBBBG is 2^{-10} and that there are ${}_{10}C_4$ such configurations containing 4G's (4 girls). Therefore the probability of exactly 4 girls is ${}_{10}C_4 2^{-10}$. Similarly we can calculate the probability of any number of heads on the toss of 10 coins, as shown in the table.

Frequency of Heads	Theoretical Probability	
0	${}_{10}C_0 \cdot 2^{-10}$	$\approx 9.76 \times 10^{-4}$
1	${}_{10}C_1 \cdot 2^{-10}$	$\approx 9.76 \times 10^{-3}$
2	${}_{10}C_2 \cdot 2^{-10}$	≈ 0.044
3	${}_{10}C_3 \cdot 2^{-10}$	≈ 0.117
4	${}_{10}C_4 \cdot 2^{-10}$	≈ 0.205
5	${}_{10}C_5 \cdot 2^{-10}$	≈ 0.246
6	${}_{10}C_6 \cdot 2^{-10}$	≈ 0.205
7	${}_{10}C_7 \cdot 2^{-10}$	≈ 0.117
8	${}_{10}C_8 \cdot 2^{-10}$	≈ 0.044
9	${}_{10}C_9 \cdot 2^{-10}$	$\approx 9.76 \times 10^{-3}$
10	${}_{10}C_{10} \cdot 2^{-10}$	$\approx 9.76 \times 10^{-4}$

The set of probabilities corresponding to each outcome is called the *binomial distribution* because the numbers, ${}_{10}C_r$ are precisely the coefficients which appear in the terms of the binomial expansion of $(1+x)^{10}$. The program, **BINOMIAL**, presented on page 43, simulates M replications of the toss of N coins. When we run **BINOMIAL** for M = 10 000 and N = 10, (that is, 10 000 tosses of 10 coins) we obtain the histogram shown below for the frequency of each of the 11 possible outcomes, i.e.: 0 heads, 1 head, 2 heads…10 heads. The experimental probabilities corresponding to these outcomes are shown in column L_3 of the table below. Observe how closely these approximate the theoretical probabilities in the table above.

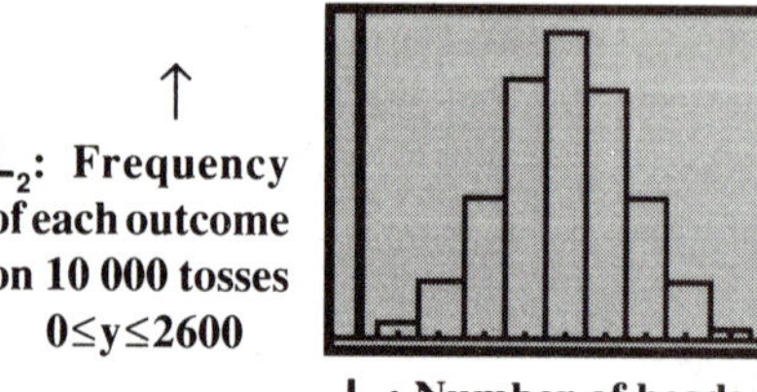

L₂: **Frequency of each outcome on 10 000 tosses** $0 \le y \le 2600$

L₁: Number of heads on the toss of 10 coins $-0.5 \le x \le 11$

We observe that the most likely number of heads is 5 and the probabilities decrease as the balance between the frequencies of heads and of tails decreases. The histogram is symmetric about the outcome "5". The area under this curve is 10 000, corresponding to the total frequency of all the outcomes. If we divide all the frequencies by 10 000, we convert the frequencies to relative frequencies, i.e. experimental probabilities. The xyLine showing experimental probabilities (L_3) vs. outcomes (L_1) is shown in the display on the right. The points on the xyLine show the experimental probabilities corresponding to each outcome. As the number of coins tossed is increased, the corresponding xyLines approach the shape of the so-called *normal curve* shown on the grid below. When this curve describes the probability densities of values of a continuous variable, it is the graph of a *normal distribution*.

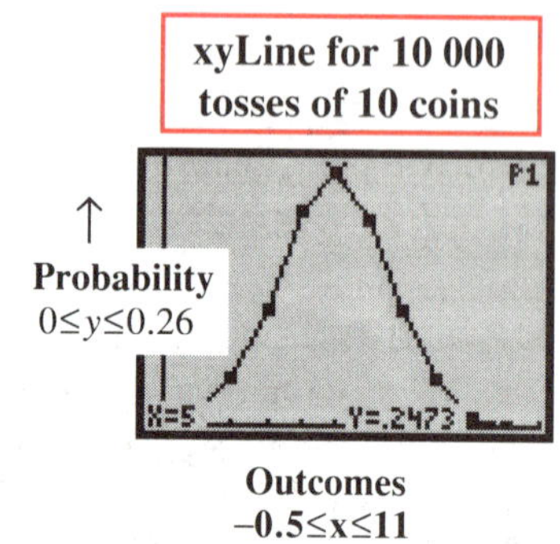

xyLine for 10 000 tosses of 10 coins

Probability $0 \le y \le 0.26$

Outcomes $-0.5 \le x \le 11$

A complete mathematical representation of the normal distribution was achieved when Abraham DeMoivre in the 1730's discovered that the equation of the curve corresponding to a normal distribution with mean, μ and standard deviation, σ is:

$$y = \frac{1}{\sigma\sqrt{2\pi}}e^{-\frac{1}{2}\left(\frac{x-\mu}{\sigma}\right)^2}$$

The constant $\frac{1}{\sigma\sqrt{2\pi}}$ was chosen so that the area under the curve is one unit.

$$y = \frac{1}{\sqrt{5\pi}}e^{-\frac{(x-5)^2}{5}}$$

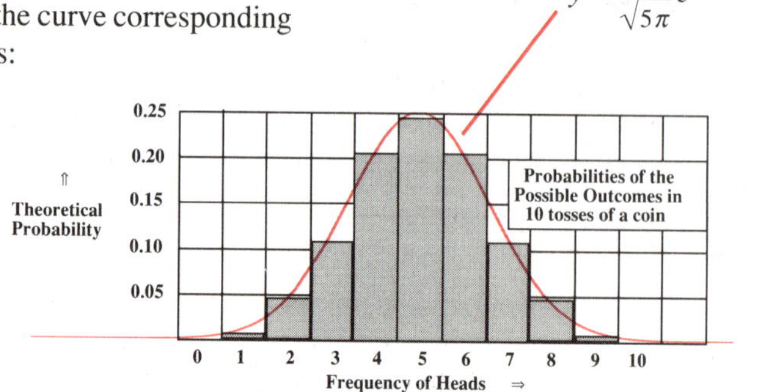

WORKED EXAMPLE

In the late 18th century, Karl F. Gauss (see Exploration 7) discovered that repeated measurements of the same object gave rise to a histogram with the general shape of the normal curve. Since Gauss was one of the important pioneers in the study of the normal distribution, it is sometimes called the *Gaussian* distribution.

In the mid 19th century, Sir Francis Galton demonstrated through extensive data collection, that many human characteristics such as height and I.Q. are distributed throughout the entire population in accordance with the normal distribution. For this reason and for its special mathematical properties, the normal distribution has become the basis of most studies involving human populations.

WORKED EXAMPLE

Galton recorded the heights of 8585 men in Great Britain. The *envelope* or outline of the histogram corresponding to these heights is a normal curve with mean, $\mu = 67$ inches and standard deviation, $\sigma = 2.5$ inches.

a) Graph the normal curve corresponding to this data and estimate the proportion of people who were between 66 and 68 inches tall.

b) What fraction of the adult male population in Great Britain was within one standard deviation of the mean height?

Solution

a) We substitute $\mu = 67$ and $\sigma = 2.5$ into the general equation for the normal distribution (given on page 50) to obtain:

$$y = \frac{\sqrt{2}}{5\sqrt{\pi}} e^{-2\left(\frac{x-67}{5}\right)^2}$$

To graph this equation, we observe that the maximum height occurs when $x = 67$, i.e. at the mean value of x. For this value of x, $y = \frac{\sqrt{2}}{5\sqrt{\pi}} \approx 0.16$. Furthermore, the heights of most men will lie between 50 and 80 inches, so we set the window to: $50 \leq x \leq 80$ and $0 \leq y \leq 0.2$. We press $\boxed{Y =}$, enter the above equation and press $\boxed{\text{GRAPH}}$ to obtain the graph in the display on the right.

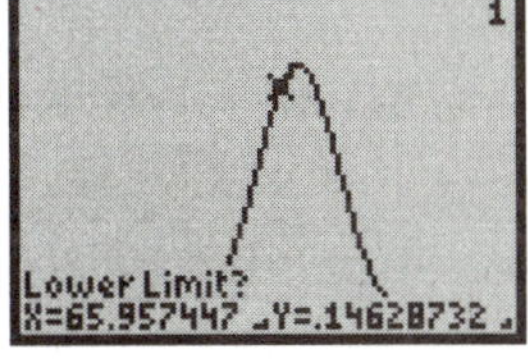

To determine the proportion of men whose heights were between 66 and 68 inches, we must find the area under the normal curve which lies in the interval: $66 \leq x \leq 68$. (This is analogous to adding the areas under the xyLine to total the probabilities that the number of outcomes lies in a particular interval.) To find this area we will use the definite integral function which is selection 7 on the $\boxed{\text{2nd}}$ [CALC] menu. If you have not studied calculus, all you need to know is that the symbol, $\int f(x)dx$, means "the area under the curve". When we press: $\boxed{\text{2nd}}$ [CALC] $\boxed{7}$ we obtain a graph and a prompt, **Lower Limit?** In response to the prompt, we move the cursor along the curve to a point as close to $x = 66$ as we can get. As shown in the display, we reach the point $x = 65.957\ldots$ To get closer to $x = 66$, we press: $\boxed{\text{ZOOM}}$ $\boxed{2}$ $\boxed{\text{ENTER}}$ and then $\boxed{\text{2nd}}$ [CALC] $\boxed{7}$ to get the display shown here.

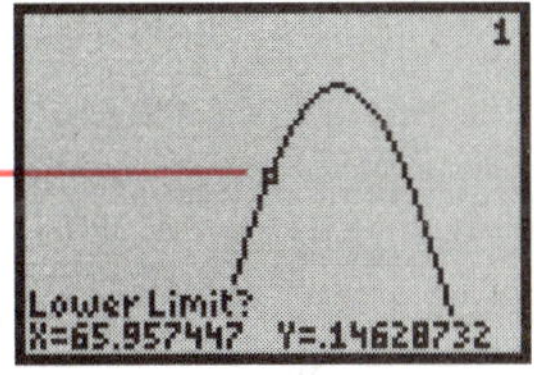

We then move the cursor to the point $x = 66.037234$ and press $\boxed{\text{ENTER}}$. In response to the prompt, **Upper Limit**, we move the cursor to the point $x = 68.031915\ldots$ and press $\boxed{\text{ENTER}}$ We obtain the display shown below right. The display indicates that the area under the normal curve and in the interval $66 \leq x \leq 68$ is about 0.310. That is, about 31.0% of the men were between 66 and 68 inches tall.

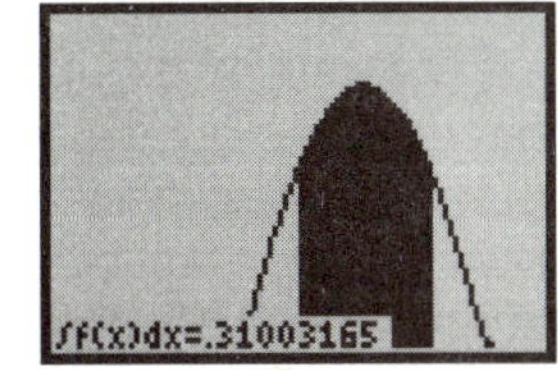

b) To determine the proportion of men whose heights lie within one standard deviation of the mean, we proceed as in part a) to determine the area under the curve between $67 - 2.5$ and $67 + 2.5$. Using the window, $60 \leq x \leq 70; 0 \leq y \leq 0.2$ we find that 68% of the men were within one standard deviation of the mean.

1. Write the equation of the normal curve corresponding to the normal distribution with:

a) $\mu = 50, \sigma = 10$ b) $\mu = 60, \sigma = 10$ c) $\mu = 80, \sigma = 10$

2. Graph each of the normal curves in exercise 1 in the range: $0 \le x \le 100$; $0 \le y \le 0.05$. How does the graph of a normal distribution change if μ is increased by k units and σ is unchanged?

3. Write the equation of the normal curve corresponding to the normal distribution with:

a) $\mu = 50, \sigma = 5$ b) $\mu = 50, \sigma = 10$ c) $\mu = 50, \sigma = 20$

4. Graph each of the normal curves in exercise 3 in the range: $0 \le x \le 100$; $0 \le y \le 0.1$. How does the graph of a normal distribution change if σ is increased by a factor, k and μ is unchanged? Describe the relationship between the dispersion of a set of normally distributed data and the shape of the corresponding normal curve.

5. a) The wavelength of the light emitted by a particular isotope of phosphorus is measured 100 times. Why might you expect the measurements to be symmetrically distributed about the mean?

    b) Prove algebraically that a normal curve is symmetric about the line x = μ.

6. The I.Q. scores of Americans on the Wechsler test of intelligence are normally distributed with a mean, $\mu = 100$ and a standard deviation, $\sigma = 15$.

a) Write the equation of the corresponding normal curve which gives the probability density corresponding to each score.

b) Graph your equation using appropriate window settings.

c) Determine the percentage of people who have I.Q. scores between: (i) 85 and 115   (ii) 90 and 100.

d) When the Wechsler test was administered to a large number of people, N of them scored 100. What score would you expect to be attained by $\frac{N}{2}$ people.

7. In a recent study of the heights of 100 000 adult females throughout the world, it was found that the height, h, is a normally distributed variable with mean, $\mu = 162.5$ cm and standard deviation, $\sigma = 6.1$ cm.

a) Write the equation of the corresponding normal curve which gives the probability density for each height.

b) Graph your equation using appropriate window settings.

c) Estimate the percentage of people who are less than 160 cm in height.

d) If 95% of all adult women are less than height, h, then h is called the 95th *percentile*. How tall is a woman who is at the 95th percentile in height?

8. In Exploration 1, we presented the histogram of SAT math scores for the males in 1990. Enter the mid points of the class intervals such as $225 \le x \le 275$, into list, L_1. (Remember to set the dimension of L_1 equal to 12.) Enter the frequencies into column L_2 and the relative frequencies into L_3. Define L_4 as **int** $(L_2/1000)$ because **1-Var Stats** requires integral frequencies ≤ 99. Define L_5 as $L_3/50$ (the class size). Plot the xyLine with Xlist L_1, Ylist L_5 and window: $200 \le x \le 800$; $0 \le y \le 0.004$.

a) Estimate the mean, μ, and standard deviation, σ, of the normal distribution corresponding to the SAT scores. (Hint: Use **1-Var Stats** with L_1 as the Xlist and L_4 as the frequency, to calculate x̄ and Sx.

b) Use the estimates of μ and σ from part "a" to write the equation of the normal curve which gives the probability densities of the SAT math scores of the males in 1990.

c) Graph this normal curve in the window: $200 \le x \le 800$; $0 \le y \le 0.004$. Display your xyLine from above together with the normal curve and compare their shapes.

d) What percent of the candidates scored within one standard deviation of the mean?

e) Estimate the score which corresponds to the 95th percentile.

9. The graph below shows the area under a normal curve with $\mu = 0$, for various intervals of width one standard deviation.

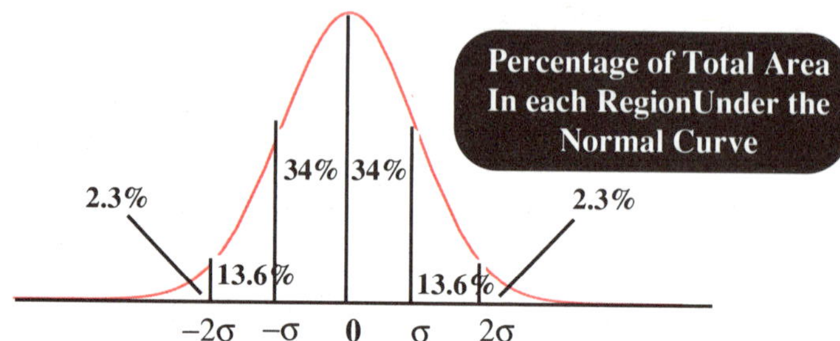

This graph shows that for *all* normal distributions no matter what mean or standard deviation, 68% of the area under the curve lies within one standard deviation of the mean. Verify this is true for all the normal curves in exercise 3.

CHALLENGE

A school board identifies a student as *gifted* if she scores more than two standard deviations above the mean on the Wechsler test of intelligence. (See exercise 6.)

    a) What score on the Wechsler test is used as the criterion for identifying a gifted student?

    b) About how many students could the board expect to have in its gifted program if there are 48,000 students in total?

Answers to Exercises
&
Hints for Investigations

"THIS IS THE PART I ALWAYS HATE."

Exploration 1

A. There are far more sober drivers than drunk drivers, so it is to be expected that sober drivers would be involved in the greater percent of accidents. A proper analysis would compare the percent of drunk drivers who have accidents with the percent of sober drivers who have accidents.

B. Dr. Wrinkles has told us that more than half of the women over 65 years of age are widows. However, we cannot draw an inference about the relative longevities of males and females from these data alone. We need additional data such as; the percent of males over 65 who are widowers and the average age gap between married women and their spouses.

C. Imagine a coin toss in which heads means rain and tails means no rain. On each of two tosses of the coin there is a 50% chance of heads and therefore one chance in four that tails will occur twice (i.e. no rain on both days). That is, the probability of rain at least once on the weekend is $\frac{3}{4}$.

D. Although 100% of all people who die of cancer breathe air, we cannot conclude that breathing air causes cancer. A proper analysis of the effects of salt would compare the percent of regular eaters of salt who acquire cancer with the percent of non-eaters of salt who acquire cancer. Even if a correlation (see Exploration 6) were discovered, a further analysis would be required to establish a causal relationship.

E. This example shows the danger in making extrapolations from a set of data. There is never a guarantee that a pattern observed in data over a particular range will manifest itself outside that range.

1. a) The death rate was 0.0017892382 or approximately 0.1789%.
  b) The death rate was 0.0016193832 or approximately 0.1619%.
  c) New York had the higher death rate for whites.
  d) New York death rate: 0.00559...%
     Richmond death rate: 0.00331...%
     Therefore New York had the higher death rate for non-whites.
  e) New York death rate from tuberculosis: 0.001862...%
     Richmond death rate from tuberculosis: 0.002240...%
  f) Richmond had the higher death rate from tuberculosis.
  g) There is an apparent discrepancy between the results of parts c and d and part f. New York had the higher death rate from tuberculosis for whites and the higher death rate for non-whites. However, Richmond had the higher death rate for people. This famous contradiction is known as *Simpson's paradox*. The paradox derives from the fact that the percentage of non-whites in Richmond is much larger than in New York. Since the non-whites have a higher incidence of death from tuberculosis, the percentage of those who die from tuberculosis is higher in Richmond. However, within each sub-population (i.e. whites and non-whites) the percentage of deaths from tuberculosis is higher in New York. As a health official, you might report that if you are a New York resident, you are more likely to die from tuberculosis than a Richmond resident of the same sub-population. However, a randomly selected resident of New York is less likely to die from tuberculosis than a randomly selected resident of Richmond (because the random selection from Richmond is more likely to turn up a non-white).

Exploration 1 (cont'd)

7. a) On average, 5000 people will have the virus and 995,000 will not have the virus.
  b) 99% of those who do have the virus plus 1% of those who do not have the virus will test positive. That is;
$$99\% \text{ of } 5000 + 1\% \text{ of } 995,000$$
$$= \quad 4950 \quad + \quad 9950$$
$$= \quad 14,900$$
  In total, 14,900 of the 1,000,000 people will test positive.
  c) The fraction of people who test positive but do *not* have the virus is: $\frac{9950}{14900}$ or 66.778...%
  d) The fraction of people who test positive and who have the virus is approximately 33.221%. The paradox here is related to Simpson's paradox. For those who have the virus and for those who do not have the virus, the test is 99% accurate. However, the group of people who do *not* have the virus is so large that most of the people who test positive are in the group for which the test is incorrect.

The Venn diagram on the right helps to illustrate the ideas. Observe that most of those who test positive do **not** have the virus even though 99% with the virus test positive and 99% without the virus test negative.

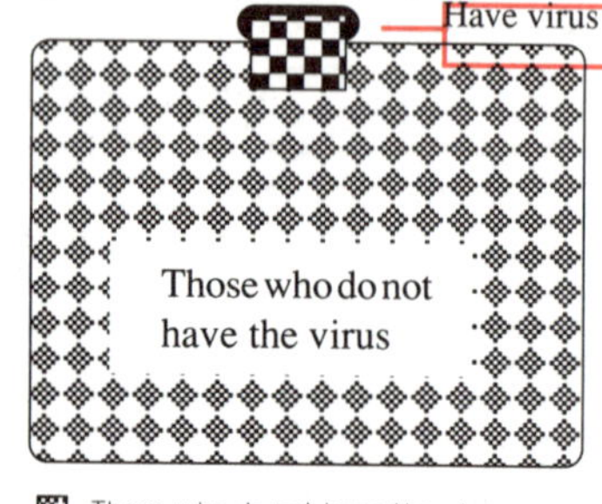

Exploration 2

1. a) 4117   b) 112,154   c) 22.87%   d) 6.53%
  e) The data in the table is grouped. That is, we know how many people scored between 700 and 750, but we do not know how many scored exactly 700. Therefore, all we can say for certain is that the number of people scoring above 700 is between 9792 and 32,039.

2. b) With the cursor positioned on L_4 of the table, we enter: $L_2 + L_3$ and we obtain the table shown below. We then define Plot1 as shown in the display on the right below.

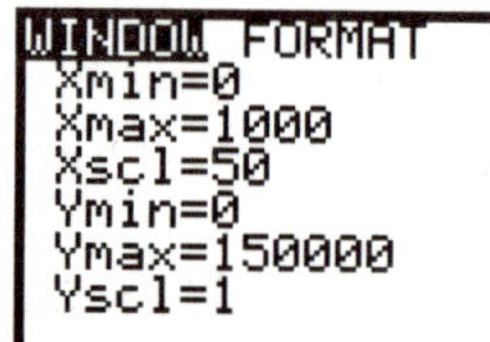

With the window settings shown on the display, we graph Plot1 to obtain the histogram shown here.

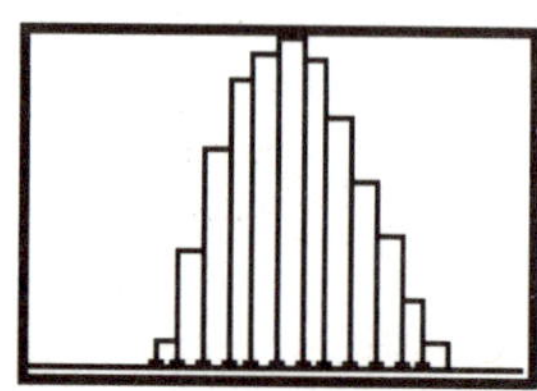

Exploration 2   (cont'd)

c) We obtain this display.

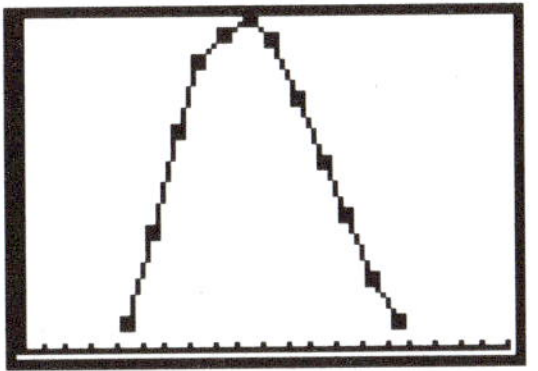

d) To change the class interval size to 100, we set Xscl to 100. Since we now have two intervals combined into one, we must increase the size of Xmax from 150000 to 300000. With the window settings on the left, we get the histogram shown on the right. By tracing along the curve to the interval $500 \le x < 600$, we discover that there are 254,176 students with scores in that interval.

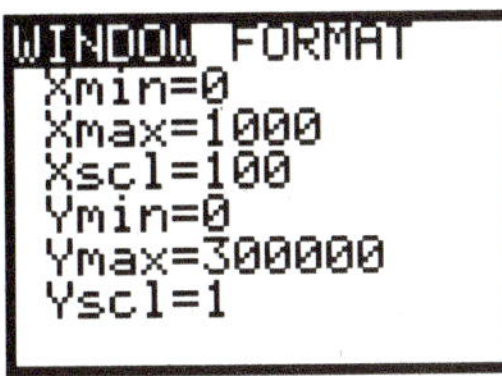
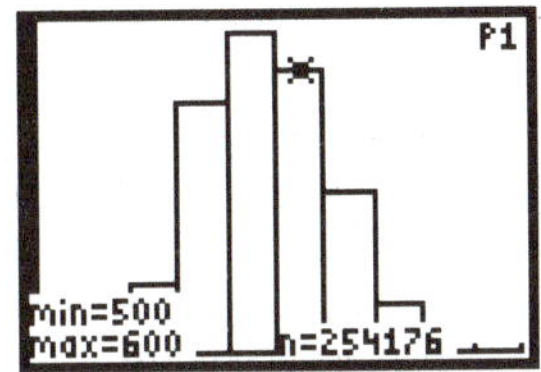

To find the total number of student scores, we evaluate **sum** L_4 using the LIST MATH submenu, and obtain the display 1025523. Therefore, 254,176 students out of 1,025,523 had scores in the 500s. That is about 24.8% of the students.

e) By tracing along the histogram to $x=600$ and keeping a running total, or by creating a cumulative frequency function as in worked example 2, we find that there were 836,711 scores below 600. This is about 81.6% of all the students.

3. a) The first 7 rows of the table are shown in the display. We then define the plot and select the window settings as indicated in the displays below.

When we graph Plot1 with these settings, we obtain the histogram shown below left. We choose xyLine, to get the frequency polygon.

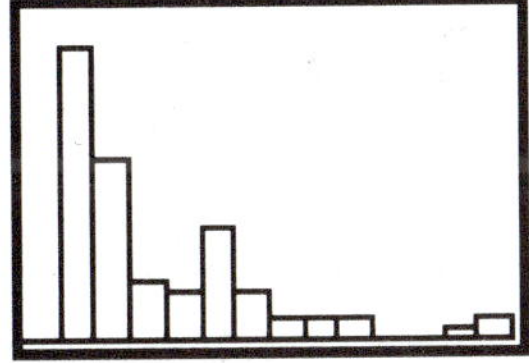

If we graph both plots simultaneously, we obtain this display. Observe that the frequency polygon joins the midpoints of the bars in the histogram.

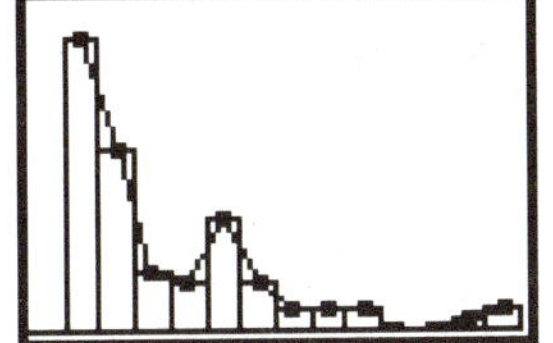

3. b) Adding the numbers in the table, we discover that there were 31 golfers out of the 87 golfers above $200,000 who won in excess of $500,000. This is about 36% of the golfers over $200,000.

c) We Xscl = 200000, Xmin = 100000 and Ymax = 60. We get the histogram shown here and trace to display 17 golfers or 13% in the interval $500,000 \le x < \$700,000$.

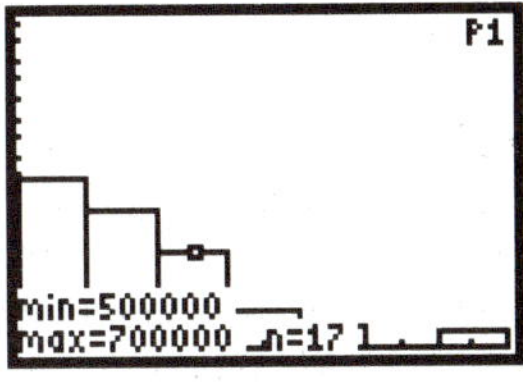

4. a) This display shows the ogive curve for the SAT scores of all students.

b) By tracing, we see that the cumulative percent of students whose scores were in the intervals up to the interval $550 \le x < 600$, (midpoint 575) was 81.58.
How could you alter your table and your window settings to obtain the ogive curve below instead of the one shown above?

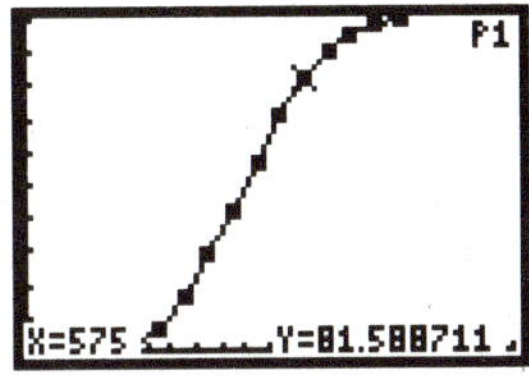
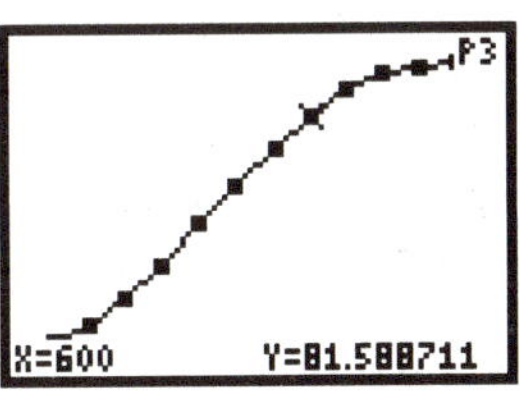

c) The ogive curve is steepest in the interval $450 \le x < 500$, because most of the student scores are in that interval, so the cumulative frequency is increasing the most in that interval.

5. a) The ogive curve is shown here. When we trace along this curve to the point $x = 500000$, the display shows that $Y = 64.3\ldots$ That is, about 64% of the golfers are below $500,000. This agrees with our result in exercise 3b).

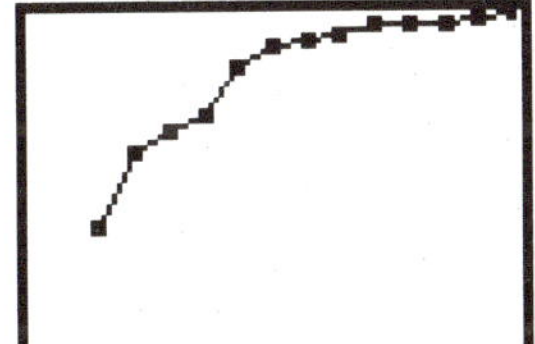
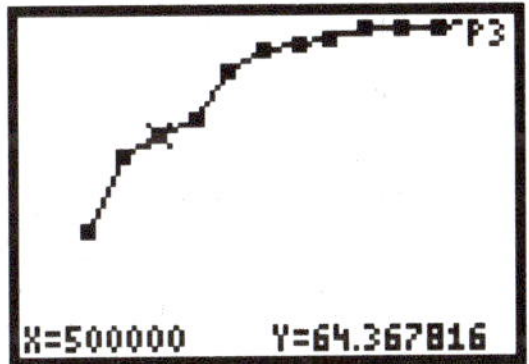

b) The curve is steadily increasing. (called *monotonic*) This is a property of all cumulative frequency curves because the height of the curve in any interval is the sum of all the heights up to and including the that interval. Furthermore, the curve is very steep for small values of x and gradually flattens out as x increases. This indicates that there are relatively few huge money winners. These are generally the golfers who win several big tournaments in the same season.

Exploration 3

1. When there is a large set of data, we seek a single value around which most of the values cluster. The particular measure of central tendency which is most informative depends upon the kind of data we have and the kind of information we seek.

2. The median would not increase at all because there are 16 employees at \$36,600 and so \$36,600 will continue to be the middle salary. However, the mean salary takes into account the value of every salary and therefore, since \$180,000 is greater than the previous mean its inclusion will increase the mean salary.

3. a) $\bar{x} = 26.416\ldots$, $Q_1 = 16$, Med = 29, $Q_3 = 34$
   b) $\bar{x} = 18.545\ldots$, $Q_1 = -17$, Med = 42, $Q_3 = 61$

4. a & b) $\bar{x} = 403.3125$, Med = 422

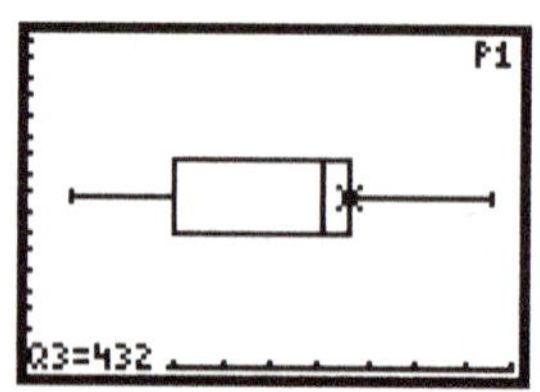

c) The quartiles are: $Q_1 = 360$, Med = 422, $Q_3 = 432$.
This can be verified by tracing along the box plot shown in the display.
The extremes are 318 and 492.
Alternatively, we could have calculated **1-Var Stats** and obtained the following displays.

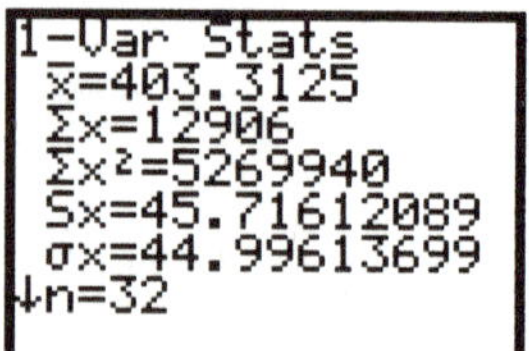

5. An *increase* of x% means that the new mean income is $(100 + x)\%$ of the original income. Therefore the new mean income was:
$0.8(172\%) + 0.2(215\%)$ or 180.6% of the original mean income. Therefore the mean increase in income was 80.6% for the entire population.
An alternative method recognizes that since the non-whites represent 20% of the population, the average increase for the entire population will be the increase for the whites plus one-fifth the difference in the increase between the whites and non-whites.
That is, $72\% + 0.2(43\%)$ or 80.6%.

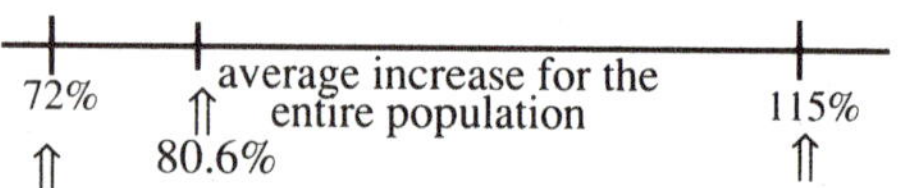

6. a) mean 50; median 50   b) mean 50.5; median $\frac{50+51}{2} = 50.5$

c) mean -1.5; median -1.5

   d) The answers to parts a, b and c suggest that the mean and median are the same for a set of numbers which form an arithmetic sequence. Is this true in general?

7. Hint: If you get an error message when you attempt to create the boxplot, round all the frequencies to the nearest thousand and record in list L_2 not the actual frequency of 11358 but its rounded frequency, 11 (thousands).

Exploration 4

1. a) The median ages of the best actor and best actress award winners are respectively 43 and 34 years.

b) Since the leading man in a movie is usually paired romantically with the leading lady, and since males have been traditionally the older member of such couples, we would expect that the set of actors from which the "best actor" is selected is older than the set of actresses from which the "best actress" is selected. Recently the age difference between males and females in such roles may be declining. If so, we might expect this difference in the age gap to be less in the second half of this quarter century.
A variety of other answers are also feasible.

c) The age range of the "best actors" is from 30 to 76 years.
The age range of the "best actresses" is from 21 to 80 years.

d) For reasons given in part b), the set of actresses from which the "best actress" is selected is younger than set of actors from which the "best actor" is selected. Hence it is more likely that the youngest best actress will be younger than the youngest best actor. However, since it takes only one person to make a new minimum, the behavior of the extreme values is much less predictable than the behavior of the median.

e) Actors and actresses who reach advanced age are less likely to be starring as couples in a romantic involvement so it would seem that the argument expressed in part b) does not apply to this age range. Another factor is the fact that women live longer than men. There would be more actresses in the advanced ages i.e. 70s, 80's and 90's than actors in these age ranges. Hence the pool from which "best actresses" are selected might be expected to have more candidates of advanced years.

f) Together, the arguments expressed in parts b) and e) suggest that the pool from which "best actresses" are selected has greater numbers at both ends of the age scale. It would seem therefore that the "best actresses" should have a larger range than the "best actors". In fact the "best actresses" have an age range of 80-21 or 59 years. The "best actors" have an age range of 76 - 30 or 46 years.

g) The interquartile range for "best actresses" is 41 - 29 or 12 years.
The interquartile range for "best actors" is 49 - 38 or 11 years.

h) The median age for winners of best actress is 9 years younger than the median age for winners of the best actor award. If there are more actresses at the advanced ages, these large values would not affect the median, but would increase the mean age of the actresses, thereby decreasing the gap in the median ages. When we use 1-Var Stats, we find the mean age for "best actresses" is 36.96 years. The mean age for "best actors" is 44.76 years. The difference in means is about 7.8 years.

2. a) 50%   b) 100%
c) Roughly speaking, when the median is the mean of the first and third quartiles.
d) Half of the values cluster closely near the median and in the remaining values, there is at least one small and one large outlier.

Exploration 5

1. Measures of dispersion were developed to determine how tightly a set of data cluster around various measures of central tendency.

2. a) Range in lifetimes for Brand A batteries: 9.3 - 3.1 or 6.2 years
b) Range in lifetimes for Brand B batteries: 7.5 - 3.3 or 4.2 years
c) The range takes into consideration only the difference between the largest and the smallest values in a set of data and ignores all the other values. Therefore, two sets of data may have the same range but significantly different clustering around the mean.

3. a) The worked example indicates that the mean is $\bar{x} = 5.7$ years. The mean deviation is $\frac{31}{30}$ or $1.03\overset{\bullet}{}$.

b) The mean deviation for Brand B batteries is $\frac{17}{30}$ or $0.5\overset{\bullet}{6}$.

c) Since Brand B batteries have a smaller mean deviation in their lifetimes, the mean life of 5.7 years is a more reliable measure for Brand B than for Brand A batteries.

4. a) $\bar{x} = 6$; $\sigma = 3.162\ldots$ b) $\bar{x} = 0$; $\sigma = 3.162\ldots$
The standard deviation of a set of data does not change when a fixed constant is added to all the values. In essence, adding a constant to all the values merely *slides* the data points horizontally without increasing their distances from the mean.

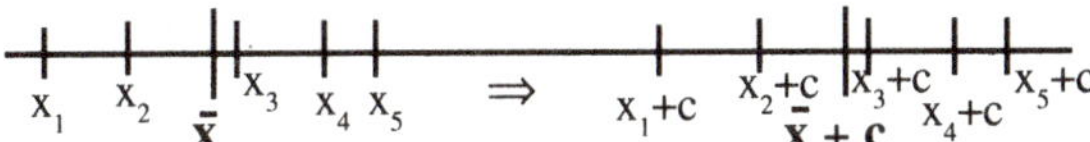

5. For the set of data, S = {1, 2, 3, 4, 5}, the mean deviation is 1.2. The standard deviation is $\sqrt{2}$. We add the data point, 81 to S, to obtain T = {1, 2, 3, 4, 5, 81}. The mean deviation of T is $\frac{65}{3}$.

The standard deviation of T is 29.097...
For the set, S, the standard deviation is about 1.178... times the mean deviation, but for T the standard deviation is about 1.34 times the mean deviation. That is, the standard deviation seems more sensitive to extreme values.

6. Diameters of 30 ball bearings from Acme: $\bar{x} = 8.0163$; $\sigma = 0.102\ldots$
Diameters of 30 ball bearings from Delta: $\bar{x} = 8.0296$; $\sigma = 0.052\ldots$
Though the mean diameter for the ball bearings from Acme is slightly closer to 8 mm than those from Delta, the standard deviation for the Delta ball bearings is about half that for Acme. That is, the Delta ball bearings have smaller deviations from the mean. Since the difference in the means is negligible relative to the difference in the standard deviations, and since most of the diameters will be within a standard deviation of the mean, the ball bearings from Delta have smaller deviations from 8 mm.

7. Histogram a) is the one which corresponds to the lifetimes of Brand B batteries. We can see this by rotating the double stem-and-leaf plot on page 20 by 90 degrees.

8. When the dispersion of a set of data is small, most of the area under the bars is concentrated near the mean value as in histogram "a" in exercise 7. For a set of data of greater dispersion, the area under the bars is distributed further away from the mean as in histogram "b".

Exploration 6

1. In general, two variables are said to be *positively correlated* if large values of each variable are associated with large values of the other variable. The degree of this correlation may be described as weak or strong depending upon how few violations of this relationship exist in the set of data points.
Two variables are said to be *negatively correlated* if large values of each variable are associated with small values of the other variable. When the correlation between two variables is approximately linear, the data points in the corresponding scatterplot cluster along a straight line and most of the data points can be contained within an eccentric ellipse with the straight line as its major axis.
When two variables are not correlated, we say that they show *no correlation*.

2. a) strong negative correlation   b) positive correlation
c) no correlation   d) negative correlation

3. The scatterplot for the ages of women and their spouses is shown below. Since the data points cluster along a straight line of positive slope, we say the ages of women are positively correlated to the ages of their spouses. Since the clustering is close to a line, we would describe this relationship as a *strong positive correlation*.

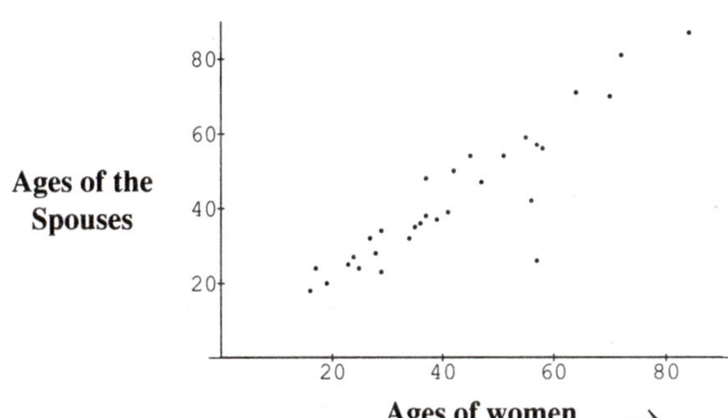

4. The scatterplot for the number of cricket chirps per minute and the outside temperature is shown below.

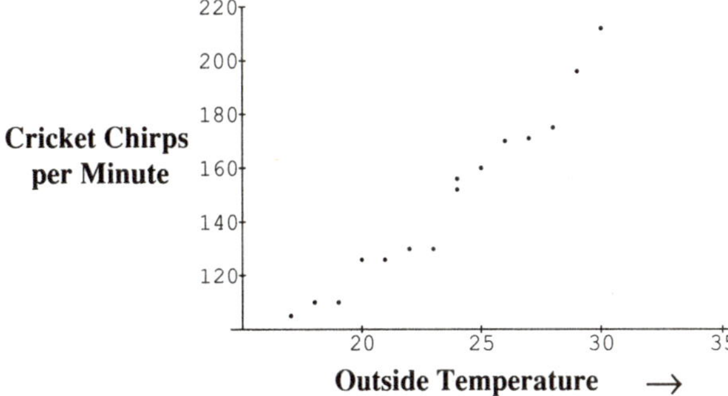

This scatterplot shows that there is a strong positive correlation between the number of cricket chirps per minute and the outside temperature. The scientist's hypothesis seems to be valid. Some might suggest that advancing such an hypothesis without using a scatterplot wouldn't be cricket. We are inclined to chirp in agreement.

Exploration 6   (Cont'd)

5. a) A positive correlation: Taller people will tend to be heavier than shorter people, so a large height is generally associated with a large weight.

b) A positive correlation: Substantial personal savings are usually associated with increased age since few young people have had the opportunity to accumulate wealth.

c) A positive correlation: The largest incomes are usually earned by people who are highly educated and the lowest incomes by people who have little or no formal education.

d) A negative correlation. Cities of large latitude are closer to the earth's poles than cities of smaller latitude and are therefore colder (lower mean temperature) in winter than cities of smaller latitude.

e) No correlation. For an individual, golf score may drop as the person improves during middle age. In older age this trend may reverse itself. However, if a data point consists of a randomly selected person's age and his (or her) golf score, then there would be little or no correlation expected, since golf scores would seem to be randomly distributed at every age level depending on experience and skill.

f) A strong negative correlation. Travel time for a particular trip is inversely proportional to the speed of travel. Since large travel times correspond to small travel speeds, these variables are negatively correlated. Since there are *no* deviations from this correspondence, the negative correlation is strong.

g) No correlation. Unless the olives are consumed with dry martinis, there is little reason to expect any correlation between these variables.

6. a) If $y = ax + b$, then $y_i = ax_i + b$ and $\bar{y} = a\bar{x} + b$, so

$y_i - \bar{y} = a(x_i - \bar{x})$. Substituting into the formula for r, we obtain:

$r = \dfrac{a}{|a|}$ and so, $|r| = 1$. That is, $r = 1$ when $y = ax + b$ and $a > 0$.

$\qquad\qquad\qquad r = -1$ when $y = ax + b$ and $a < 0$.

b) Since there is no universally agreed range of r corresponding to the terms *strong correlation* and *weak correlation*, we show the general ranges on a spectrum.

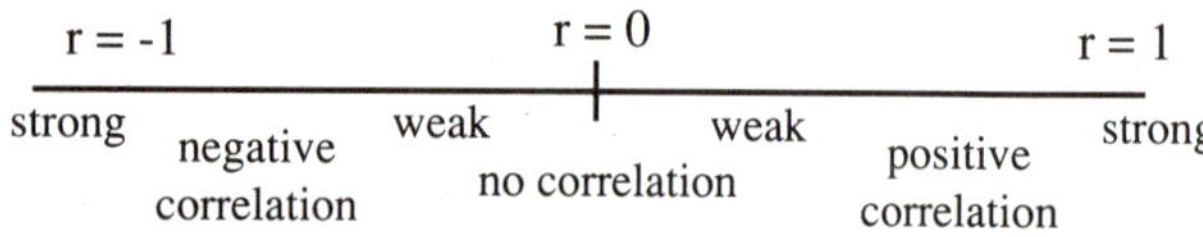

The (linear) correlation coefficient between the number of cricket chirps per minute and the outside temperature is: $r = 0.9768\ldots$

7. a) The (linear) correlation coefficient between the age of a woman and the age of her spouse is: $r = 0.90716\ldots$

b) The (linear) correlation coefficient between the hours of study and the test mark is $r = 0.7650\ldots$

Note: Compare these coefficients with the correlation of 0.4 between the IQ's of parents and their children, or the correlation of 0.5 between the intelligence of siblings.

Exploration 7

2. Following the step-by-step instructions, we divide the measurements into three groups: 1928-1952; 1956-1972 and 1976-1992. The summary points for these groups are respectively, (1936, 10.30), (1964, 10.14) and (1984, 9.99). We join the first and third summary points with a line segment and then draw a parallel line through the middle summary point. We then slide a ruler one-third the distance from the line segment to the line and trace the median-median line. To do this accurately, extend the line segment to where it cuts the y-axis and then slide one-third the distance up the y-axis to the y-intercept of the line.

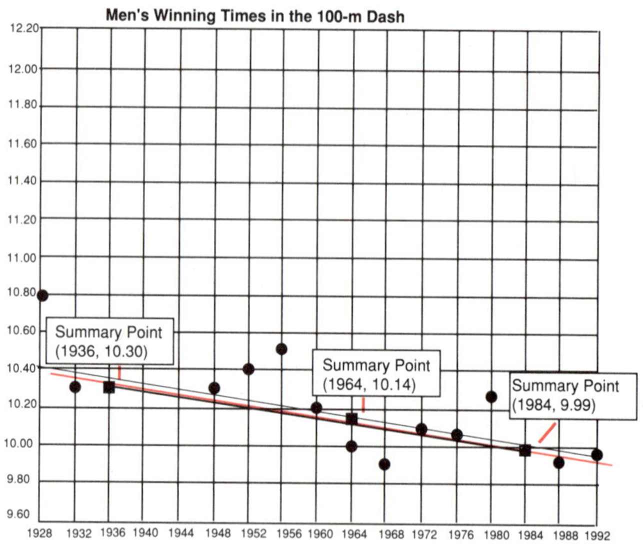

When we construct the median-median line on our calculator as in the worked example, we obtain the display shown below. Using the STAT CALC menu, we obtain the equation of the median-median line indicating that its slope is -0.006 and y-intercept 22.810. Note that the slope of the median-median line is the same as the slope of the line segment joining the first and third summary points. Therefore we can calculate the slope of the median-median line by calculating the slope of the line segment.

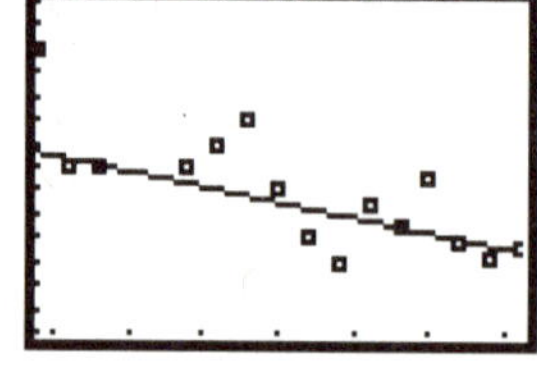

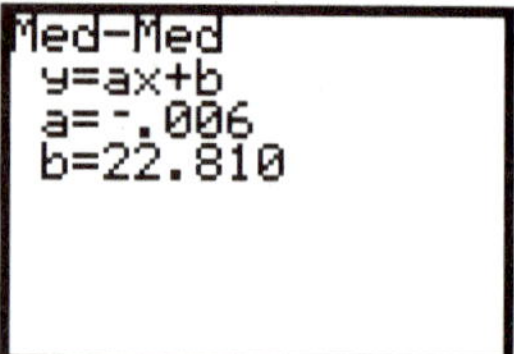

Note: The median-median line is an effective device for smoothing out fluctuations in data and looking for long term trends, in such applications as sale analysis, stock market indices and monetary exchange rates. However, extreme caution must be exercised when using the median-median line to make predictions. The median-median line approach is based on the assumption that the underlying relationship between the variables is linear. If this assumption is incorrect (and it usually is incorrect) the predictions have no validity.

Exploration 8

1. The method of least squares is a method for finding the equation of a particular curve in a family of curves (such as linear, logarithmic, exponential etc.) for which the squares of the vertical distances of the data points from that curve is a minimum. This curve is called the *regression line*. When seeking a straight line (called the linear model), we choose the straight line of best fit and it is called the *regression line of y on x*. However, if we are seeking an exponential or logarithmic curve to approximate our data, the line of best fit is referred to respectively as the *exponential* or the *logarithmic* regression line of y on x.

The residual of a data point is the observed value of the dependent variable minus its predicted value (obtained by substituting the value of the independent variable into the equation of the regression line).

2. If both axes have the same scale then the slopes are as follows:
a) slope ≈ -0.8   b) slope ≈ 0.8   c) slope ≈ -0.6

3. Both the median-median line and the regression line are straight lines which attempt to describe a linear relationship between the independent and dependent variables. They both assume an underlying linear relationship between the two variables and are used to smooth out data and predict values of one variable from the other.
The differences between a median-median line and a regression line are similar to the differences between a median and a mean. Substantial increases or decreases in the extreme values of a set of data will have no effect on the median-median line but it will affect substantially the equation of the regression line. That is, outliers have minimal effect on a median-median line but play a significant role in determining a regression line.

4. The curve of best fit:
in the logarithmic model, has equation $Y = a + b \ln X$
in the exponential model, has equation $Y = ab^x$
in the power model, has equation $Y = ax^b$
where "a" and "b" are parameters.

5. To compare regression curves of various types, we find the regression equation for various models, such as; linear, logarithmic, exponential etc. and calculate the residuals as in the worked example on page 29. The curve for which the sum of the squares of the residuals is smallest is the curve which we select as the regression curve.

6. a) If the relationship were linear, increasing the hours of study by one hour should increase the mark by a fixed amount no matter what mark a student has. However, a student with a high mark say, 99, cannot improve the exam mark no matter how many hours she works. This is called *the law of diminishing returns*.

  b) The law of diminishing returns suggests a logarithmic relationship. The opportunity to improve a mark is greatest when the study hours are small. Once a student has invested large numbers of hours in study, the mark is so high that there is little room for improvement; i.e. increasing *x* by a lot, generates small growth in y.

7. a & b) The scatterplot is shown below along with the graphs and equations of the linear and logarithmic regression lines.

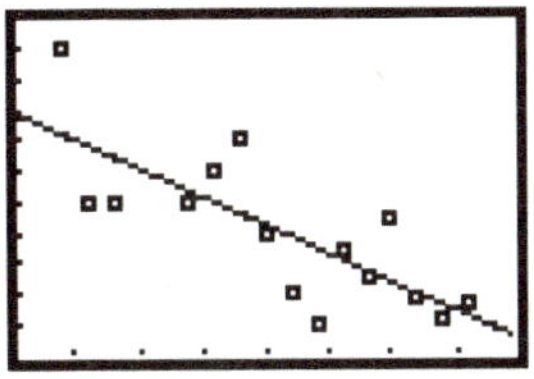
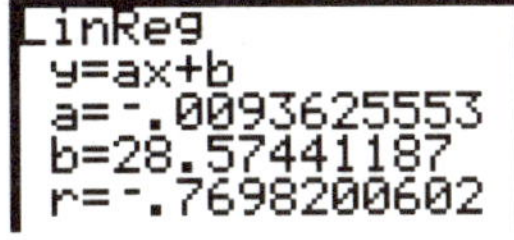

Observe that only one regression line is visible because they are so close that they are almost coincident.

To examine each line in turn, merely deselect one of the curves on the Y= list.

c) To determine which regression line gives a better fit, we calculate the sums of the squares of the residuals for both curves, following the procedures demonstrated in the worked example, page 29. The sum of the squares of the residuals for these two regression lines are 2.56 and 2.54 for the linear and logarithmic lines respectively. It seems that the logarithmic model is a slightly closer match. When we trace along each curve in turn, we find the prediction of about 9.38 seconds for 2052 using the logarithmic regression line. The median-median line predicted a winning time of 9.56 seconds.

Note: This display shows in order from top to bottom the exponential, logarithmic and linear regression lines in the interval: 2035≤x≤2070
9.13 ≤y ≤9.60

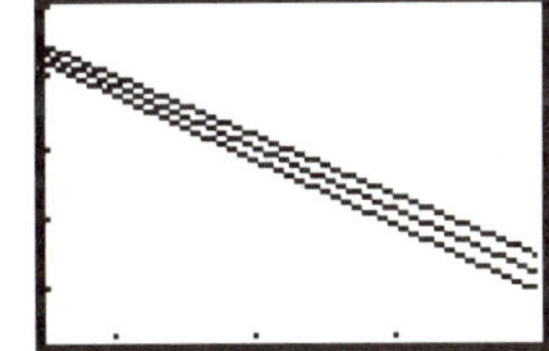

Since the linear model predicts a 0 winning time eventually, it is not reliable too far outside the data range.

8. The scatterplots and two regression lines are given in the display.

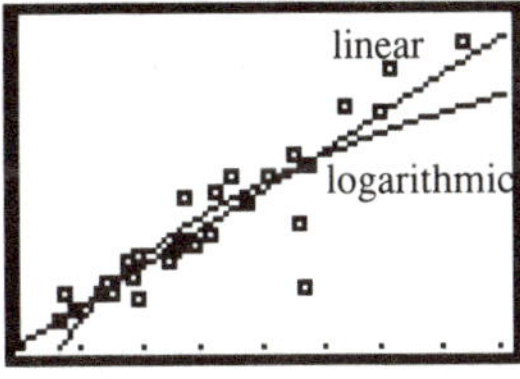

a) r = 0.90716... This is a strong correlation.
  Equation of linear regression line is: Y= 2.811...+ 0.94869...X

b) r = 0.8699...The linear regression line provides a better match to the data because the sums of the squares of the residuals is smaller than for the logarithmic regression line.
  Equation of logarithmic line is: Y = -90.264...+ 36.421...lnX

If men are generally *x* years older than their wives at marriage, then Man's age = Spouse's age + x. That is, we have a linear relationship. At advanced ages women's greater longevity could affect this unless the age of dead spouse is contained in the data.

c) The estimated age of the husband of a 48-year-old using linear regression is about 48.34 years.

Exploration 9

1. Following the procedures in the worked example, we find:
a) 4   b) 12   c) 24   d) 24   e) 5   f) 20   g) 60   h) 120

2. a) The correct formula is $_nP_r = \dfrac{n!}{(n-r)!}$ for $r \leq n$.

b) $_nP_r$ is the number of permutations of n items taken r at a time. We can create all such permutations by the following process. Choose the first item in n ways and for each of these ways… Choose the second item in n-1 ways, and for each of these,… Choose the third item in n-2 ways, and for each of these,…

• • • • • •

Choose the $(r-1)^{th}$ item in $n-r+2$ ways and for each way … Choose the rth item in $n-r+1$ ways.

It follows from the Fundamental Counting Principle (page 16) that the total number of ways of choosing r items from n items is:
$$n(n-1)(n-2)\ldots(n-r+1) \quad \text{or} \quad \frac{n!}{(n-r)!}$$

If you find difficulty following the foregoing argument, apply it to $_{10}P_3$ in the worked example. For each of the $_{10}P_3$ ways for the 1st, 2nd and 3rd horses to finish, there are 7! ways for the remaining horses , yielding a total of $_{10}P_3 \cdot 7!$ possible outcomes.

c) When n = r, the formula reduces to n! because 0! = 1.

3. There are $_{26}P_6$ permutations of 26 letters taken 6 at a time.
$_{26}P_6 = 165,765,600$

4. If the year of your birth has repeated digits, then the probability that the security code will be the year of your birth is zero.
The total number of 4-digit security codes containing four different digits is $_{10}P_4$ or 5040. Therefore if your birth year has no repeated digits, the probability the security code will be the year of your birth is one chance in 5040 or 1.9841×10^{-4}.
Note that the probability that the 4-digit code is *your* birthday is not the same as for a randomly chosen birthday.

5. Reasoning as in the worked example, we find the probability is:
$$_9P_3 = \frac{1}{504} \text{ or about, } 0.00198\ldots\text{which is approximately } 0.2\%.$$

Exploration 10

1. Using the procedure in worked example 1, we compute the following:
a) $_8C_4 = 70$ b) $_{12}C_{10} = 66$ c) $_{21}C_{15} = 54265$ d) $(_{12}C_{10})^{-1} = 0.015$

2. a) $_9P_4 = 3024$; $_9C_4 = 126$; $_9P_4/_9C_4 = 24$ or 4!

b) We enter $_{17}P_{12}/_{17}C_4$ and obtain 479001600
We evaluate 12! and obtain: 479001600

3. Since the order of the cards in the hand is not relevant, we want the number of ways of choosing 5 cards from 52 or $_{52}C_5 = 2{,}598{,}960$

4. a) $_{20}C_3$ or 1140   b) 3! or 6   c) $6 \times 1140 = 6840$

5. Choose the first digits 3 ways and for each way, the last four in 10^4 ways. That is, 30,000 different numbers altogether.

Exploration 10 (cont'd)

6. There are $_{14}C_2$ ways of choosing 2 teams form 14, so there is one chance in $_{14}C_2$ or 1 chance in 91. The probability is 91^{-1} or 0.0109…

7. a) Represent a girl by G and a boy by B. The first child can be either a B or a G and for each of these ways the second child's gender can be either a B or a G, and for each of these ways…and so on. In all there are 2^4 or 16 possible gender configurations.
How many of these gender configurations have exactly 2 Gs?
Of the 4 positions: eldest, second eldest, third eldest, and youngest, there are $_4C_2$ ways of assigning these to girls.
Therefore the probability is $_4C_2/16$ or 6/16 = 0.375.

b) The tree diagram shows all 16 possible cases. Of these, the 6 cases which have two Gs are contained in boxes.

8. We can represent a birth configuration as a 7-digit permutation such as; GGBBGBG, which indicates that the first born was a girl, the second born a girl, the third born a boy, etc. To determine the number of configurations with exactly 4 boys, we must count how many ways we can assign 4B's to positions in the set {first, second, third, …seventh}. There are $_7C_4$ ways of selecting 4 out of 7 possible positions. That is, there are $_7C_4$ equally probable configurations of exactly 4 boys. Furthermore, there are 2^7 possible configurations (i.e. 2 ways of choosing B or G for each of the 7 positions).

The probability of a configuration of exactly 4 boys is $\dfrac{_7C_4}{2^7} = 0.273$.

9. a) Reasoning as in exercise 4, we consider each possible answer sequence as a 10-digit permutation of C's (correct answer) and I's (incorrect answer). For example, CCICIICCII (though it looks like a Roman numeral) corresponds to a student answer sheet in which correct responses were given to items 1, 2, 4, 7 and 8.
The number of 10-digit permutations of C's and I's which contain 7 C's is $_{10}C_7$. The number of possible sequences is 2^{10}. Therefore, the probability that the answer sequence contains exactly 7 C's is: $\dfrac{_{10}C_7}{2^{10}}$ or 0.11718.

b) Probability of *at least* 7 correct = $\dfrac{_{10}C_7 + _{10}C_8 + _{10}C_9 + _{10}C_{10}}{2^{10}}$
which reduces to 0.1718…, so there is about one chance in 6.

10. a)   (i) $_{17}C_{12} = 6188$
(ii) $_9C_6 * _8C_6 = 2352$
(iii) $_9C_6 * _8C_6 + _9C_5 * _8C_7 + _9C_4 * _8C_8 = 3486$

b) Number of ways of choosing the committee with exactly 6 boys is 2352 (from (ii)). Number of possible committees is 6188 (from (i)). Probability is 2352/6188 = 0.38….

Exploration 10 (cont'd)

11. A *triactor box* requires that the bettor select the 3 top horses out of the field of 9. There are $_9C_3$ ways of selecting 3 horses from 9. Therefore, the probability that the correct selection will be made is: $(_9C_3)^{-1}$ or one chance in 84, which is 0.0119…That is, there is slightly better than a one in a hundred chance of winning the triactor in a random box.

12. As observed in the answer to exercise 4 of Exploration 9, if the year of your birth has repeated digits, then the probability that the security code will be the year of your birth is zero. However, if the digits do not repeat, then your birth year has four *different* digits and we can reason as follows:
The number of ways of choosing 4 different digits out of 10 is $_{10}C_4$
The number of permutations of each set of 4 digits is: 4!
Therefore, the total number of 4-digit codes possible is: $_{10}C_4 \cdot 4!$
The probability that a randomly chosen 4-digit code is identical to your birth year is: $(_{10}C_4 \cdot 4!)^{-1}$ or 1.9841×10^{-4} as in exercise 4 of Exploration 9.

13. a) A pentagon has 5 vertices. There are $_5C_2$ ways of pairing or joining these vertices. Five of these joinings are sides and the rest are diagonals. Therefore the number of diagonals is $_5C_2 - 5$ or 5.

b) Reasoning as in part a, we conclude that the number of diagonals of a hexagon is $_6C_2 - 6$ or 9.

c) Reasoning as in parts a and b, we conclude that the number of diagonals of an octagon is $_8C_2 - 8$ or 20.

d) Generalizing from parts a, b and c above, we conclude that the number of diagonals of a polygon of n sides is $_nC_2 - n$ or $\frac{n!}{(n-2)!2!}$.

That is, a polygon of n sides has $\frac{n^2-3n}{2}$ diagonals.

14. There are $_nC_2$ possible pairs of mathematicians in a group of n mathematicians. Hence, the number of handshakes was $_nC_2$ or $\frac{n(n-1)}{2}$.

Exploration 11

1. The theoretical probability of an event is the result of a computation based on the assumption that a certain set of outcomes are equilikely. The experimental probability of an event is an estimate of its theoretical probability based on the observed relative frequency of that event on a large number of replications.

2. Since there is no reasonable way to compute the theoretical probability of each outcome, we compute the experimental probability by tossing the cup a large number of times and using the relative frequency of each outcome as the experimental probability.

3. a) The U.S. Department of Health and Human Services has reported that in 1989, about 934,000 Americans died from heart disease. This represented approximately 43% of all American deaths in that year. Assuming that the basic factors do not change from year to year, we take this as an estimate of the proportion (relative frequency) of deaths from heart disease which will occur in future years.

Exploration 11 (cont'd)

3. b) This probability is based on observation and is therefore an experimental probability. Causes of death are too complex to permit the creation of a simple model which would permit the computation of a theoretical probability.

Since the answers to the exercises and investigations which follow are based upon simulations, they will vary within a range. Consequently we will give the theoretical probabilities where known, so that you can compare your results to what is predicted. If your answers differ significantly from the theoretical probability given, repeat your simulation for a larger number of replications.

4. a) The theoretical probability of heads is 0.5, so the expected frequency is about 50. However, the theoretical probability of *exactly* 50 heads is: $_{100}C_{50}/2^{10}$ or 0.0795, so it is not likely you will get exactly 50 heads. However, the probability that the number of heads, H, satisfies $45 \le H \le 55$ is about 0.73. Therefore, the frequency of heads should be 50 ± 5 about 3 times out of 4.

b) On each of the 10 replications of the simulation, it is just as likely that the frequency of heads will fall short of 50 as exceed 50. Furthermore, from part a, we expect that most of the 10 frequencies will fall between 45 and 55, so the median should be close to 50.

5. Answers will vary somewhat, but the relative frequency of heads will *usually* approach closer to one-half as the number of tosses increases. In a small proportion of simulations there will be exactly 25 heads on 50 tosses, so it will not be possible in such cases that 1000 tosses will yield a relative frequency which is closer to one-half. However, since the dispersion of the set of relative frequencies decreases as the number of tosses increases, we can expect the relative frequency of heads to approach closer to one-half as the number of tosses increases.

6. a) The probability that A will win is $\frac{3}{4}$, so we would expect that A will win about 75 of the 100 games. The probability that A will win *exactly* 75 of the games is about 0.0918; however the probability that A will win between 70 and 80 games (i.e. $70 \le x \le 80$) is about 0.78.

b) The median number of wins on the 10 replications should be close to 75.

c) The results in part b will probably indicate that the median number of wins for A is closer to 75 than to 66, so the Pascal model seems to fit observations more closely than the Roberval model.

7. Most people expect the number of heads and tails to balance and they expect H - T to approach 0. However, theoretical probability that H - T will be exactly 0 after 100 tosses is about 0.056. It is the expression, $\frac{H-T}{N}$, which approaches 0 as N becomes very large.

9. a) 0.5
b) 0.33…
c) 0.833…
d) 0.33…

Exploration 12

1. A simulation is a method for obtaining the experimental probabilities for various outcomes of an experiment by running a large number of replications of a similar statistical experiment whose outcomes should have the same probabilities as those we wish to measure. By observing the relative frequencies of the outcomes in the simulation, we can estimate the probabilities we seek to measure.

2. There is no way we can ever be sure that an experiment is behaving in a typical way and so we never know when an experimental probability is a close approximation to the corresponding theoretical probability. However, as we increase the number of replications of the simulation, we reduce the chances that such an aberration will occur.

Since the answers to the exercises and investigations which follow are based upon simulations, they will vary within a range. Consequently we will give the theoretical probabilities where known, so that you can compare your results to what is predicted. If your answers differ significantly from the theoretical probability given, repeat your simulation for a larger number of replications.

3. Since a correct and an incorrect guess on an item are equilikely, we can use COINTOSS to simulate guessing. We can associate with the outcome, "heads", the outcome of selecting a correct answer.

    a) The theoretical probability of exactly 7 correct responses is: $\frac{_{10}C_7}{2^{10}}$ or 0.117.

    b) The theoretical probability of at least 7 correct is 0.1718… (See exercise 9 of Exploration 10 for more detail.) Our simulation should produce at least 7 heads on about 17% of the simulations and this estimate should become more reliable as the number of replications is increased.

4. Since both genders are equally likely at birth, we can use the COINTOSS simulation and identify "heads" with the occurrence of a girl.

    a) The theoretical probability of exactly 4 girls is $\frac{_6C_4}{2^6}$ or 0.234..

    b) The theoretical probability of at least 4 girls is:
$$\frac{_6C_4 + _6C_5 + _6C_6}{2^6} \text{ or } 0.343\ldots$$
The COINTOSS simulation should yield at least 4 heads on about one-third of the replications.

5. Since each item offers 6 equally likely choices, we can simulate the selection of an answer with the toss of a die. We identify rolling a ⊡ with the selection of a correct answer. The process of selecting 20 answers can be simulated by running DIETOSS for 20 tosses. The theoretical probability that the student will *not* choose correctly on at least one-quarter of the items, is about 0.768749. That is, the theoretical probability that the student will choose the correct answer on at least 5 of the items is 0.2312… The DIETOSS simulation should yield at least 5 correct responses on about 23% of the replications.

Exploration 12   (Cont'd)

6. The probability that a female is hired for a particular job in that occupation is one-third. Therefore, we can simulate the hiring of 82 employees for Acme by 82 tosses of a die. We identify the occurrence of a 1 or a 2 on the die with the hiring of a female and any other outcome with the hiring of a male. We run DIETOSS with N = 82 and record the total number of tosses on which a 1 or 2 occurs.

Repeat this process about 20 times (It takes a little while to run.) and for each replication, add the two top numbers in column L_2 of the STAT EDIT table displayed by the simulation. (This is the number of times a 1 or 2 has appeared on the die.) Record the number of replications for which this total was 21 or less.

The theoretical probability that 21 or fewer females would be hired as a result of random selection, is about 0.052. That is, only one or two replications out of 20 should yield a total of 21 or less.

This analysis is not enough to prove discrimination, since all of the employees may have been hired at a time when the percentage of women in the work force was much smaller. Many factors would need to be considered before inferences could be drawn.

7. Since there is one chance in 6 that a passenger will not show, we can simulate the 96 decisions of the booked passengers by 96 tosses of a die and we can identify the outcome ⊡ with the decision not to show. We run DIETOSS simulation for 96 tosses and record whether the outcome ⊡ occurs fewer than 11 times, creating a seat shortage. We repeat this simulation a total of 20 times and record the fraction of simulations on which there are fewer than 11 occurrences of ⊡. This gives us an estimate of the probability that more than 85 of the 96 booked passengers will show.

The theoretical probability that more than 85 passengers will show is:
$$_{96}C_0\left(\frac{5}{6}\right)^{96} + _{96}C_1\left(\frac{1}{6}\right)\left(\frac{5}{6}\right)^{95} + \cdots + _{96}C_{10}\left(\frac{1}{6}\right)^{10}\left(\frac{5}{6}\right)^{86} \approx 0.06$$

There is only about a 6% chance that there will be a seat shortage.

8. This problem is similar to worked example 2, but it can be simulated with a single die since there are only 6 positions. We run the simulation DIETOSS, for 20 tosses and record whether there is at least one zero in the matrix displayed. If not, all 6 outcomes have occurred at least once. That is, the player has been in all 6 positions at least once. We call this the "non-zero" outcome. To estimate the probability, we repeat the simulation 30 times and record the relative frequency of the non-zero outcome over the 30 replications.

The theoretical probability that a particular member of the Old Timers will have played all positions after 20 games is about 0.85.

9. b) The actual areas are:

  (i) $\pi\sqrt{14}$ (ii) $3\sqrt{3}\pi$ (iii) $\dfrac{97\pi}{\sqrt{84}}$

10. a) When we ran ELLIPSE with A = √2 and B = √7 and N=1000, we obtained these displays: (Remember:press ENTER when the points have finished plotting.) Estimated Area 0.779(4√14)=3.116√14

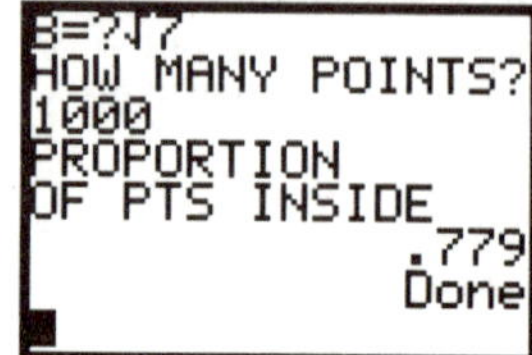

Exploration 12 (Cont'd)

10. b) The program, HCYCLOID is shown here.

c) When we ran HCYCLOID for 100 points, we got these displays.

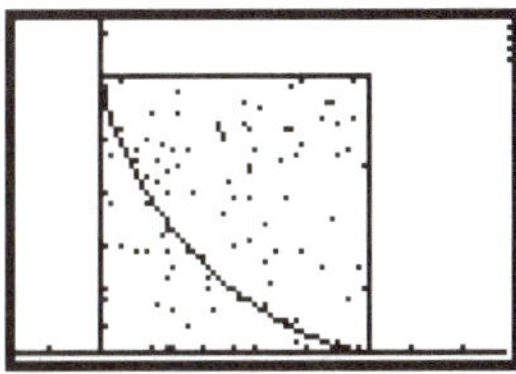

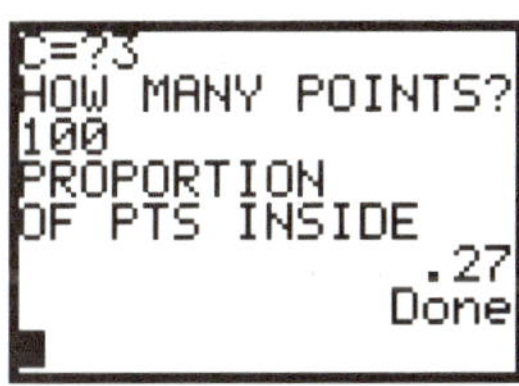

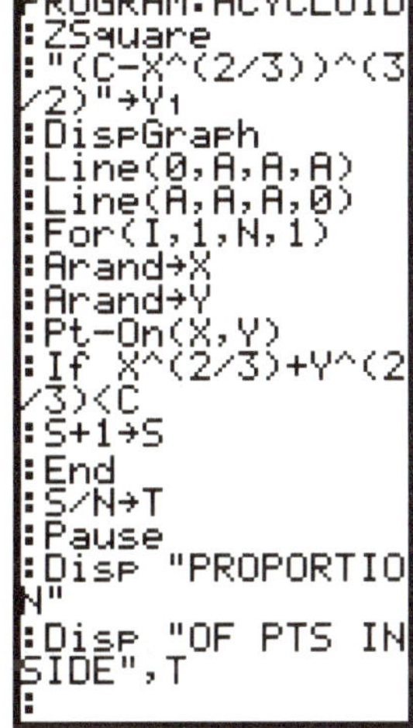

Formula for area under a hypocycloid with c = 3 is: $(3/32)\pi c^3$ or $7.95\ldots$. The simulation indicates that the area under the hypocycloid is 27% of the area of the square. But the square has side length $3\sqrt{3}$ and therefore area 27 square units. The area under the hypocycloid is therefore approximately 27% of 27 or 7.29 square units.

d) The area is given by the formula $A = k\pi c^3$. On solving for k we get $k = 0.085\ldots$
The actual value for k is 3/32 or $0.0937\ldots$
If we increased our simulation to 1000 points we would probably obtain an estimated value of k closer to the true value.

11. a) When we ran BINOMIAL for 10,000 tosses, we obtained the histogram in the display showing the frequency of each of the 8 possible outcomes; i.e. 0 boys, 1 boy, 2 boys……7 boys. The table shows the frequency (L_2) of each outcome (L_1) and the relative frequency (L_3) of each outcome.

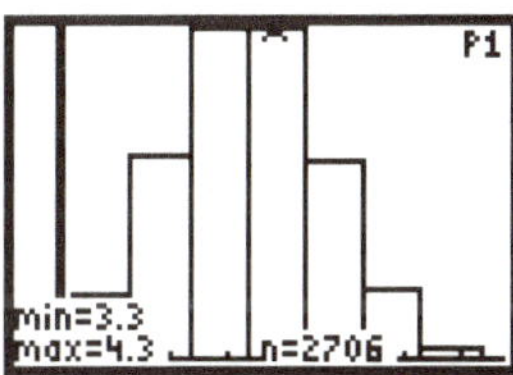

L_1	L_2	L_3
0	78	.0078
1	543	.0543
2	1660	.166
3	2722	.2722
4	2706	.2706
5	1619	.1619
6	569	.0569
7	103	.0103

$L_1(9)=$

b) In exercise 8 of Exploration 10, we calculated the theoretical probability of exactly 4 boys to be 0.273. Running the BINOMIAL simulation, we obtained (See table above) an estimate of 0.2706. This shows close agreement between theoretical and experimental probabilities.

12. See worked example 1 on page 46.

Exploration 13

3. The theoretical probability is given by:

$$1- [0.8^{20} +20(0.2)(0.8)^{19}+ {}_{20}C_2(0.2)^2(0.8)^{18}+ {}_{20}C_3(0.2)^3(0.8)^{17}$$

$$= 0.58855\ldots$$

5. In factorial notation, the expression is: $\dfrac{365!}{342!(366)^{23}}$

We can evaluate the quotient 365!/342! by evaluating ${}_{365}C_{342}$ and then multiplying the result by 23!. We then divide by the power 366^{23}, and the display shows $0.4626\ldots$
However, if we tried to evaluate 365!, we would get an overflow error.

6. b) The theoretical probability is: $1 - \left(\dfrac{11}{12}\right)\left(\dfrac{10}{12}\right)\left(\dfrac{9}{12}\right)\left(\dfrac{8}{12}\right)$

which is equal to $0.618\ldots$

8. We simulate with a single die, the six 5-minute intervals in which the newspaper may arrive. We simulate with another die, the six 5-minute intervals in which Professor Polyhedron during which she leaves the house. The two five minute intervals in which they overlap correspond to outcomes (5, 1), (5, 2), (6, 1) and (6, 2) on the toss of two dice. We run TWODICE for 5 tosses (one day out of 5) and check the four elements in the lower left corner of the matrix for a non-zero entry. We then repeat the simulation 20 times.
The theoretical probability that she receives the paper is about 0.45.

12. When we ran BINOMIAL for 10000 tosses of 10 coins, we obtained the histogram shown in this display

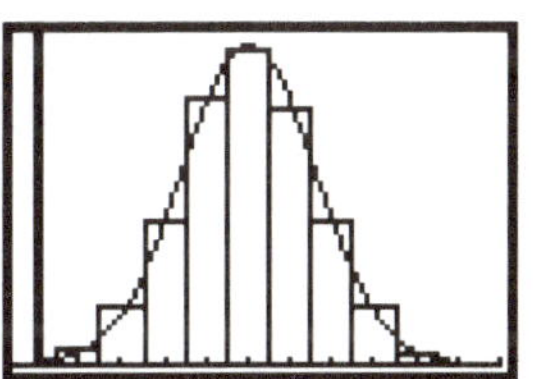

Superimposed curve has equation
$y= 10000/(\sqrt{(5\pi)})e\verb|^|(-(x-5)^2/5)$

When we ran BINOMIAL for 10000 tosses of 20 coins, we obtained the following table of frequencies and relative frequencies for the 21 possible number of heads 0, 1, … 20.

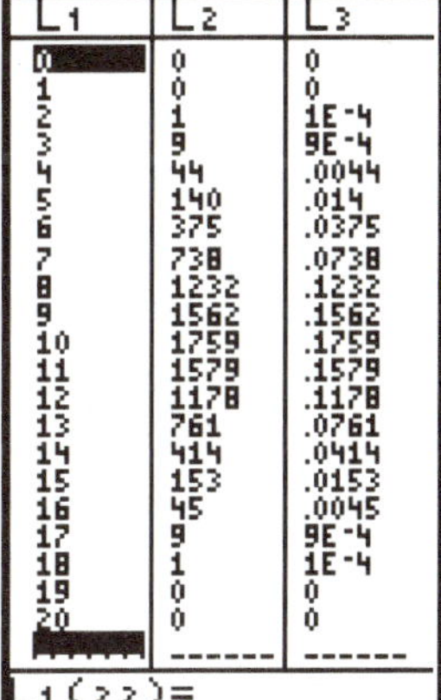

L_1	L_2	L_3
0	0	0
1	0	0
2	1	1E-4
3	9	9E-4
4	44	.0044
5	140	.014
6	375	.0375
7	738	.0738
8	1232	.1232
9	1562	.1562
10	1759	.1759
11	1579	.1579
12	1178	.1178
13	761	.0761
14	414	.0414
15	153	.0153
16	45	.0045
17	9	9E-4
18	1	1E-4
19	0	0
20	0	0

$L_1(22)=$

The displays below show the histogram of the frequencies of the outcomes and superimposed on the histogram, the curve with equation

$y=10000(\sqrt{10\pi})^{-1}e\verb|^|((x-10)^2/10)$

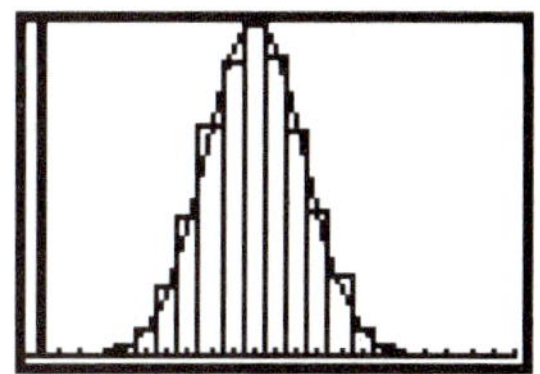

Exploration 14

Exploration 14

1. The equations are: $y =$

a) $\dfrac{0.1}{\sqrt{2\pi}}\,e^{-\frac{1}{2}\left(\frac{x-50}{10}\right)^2}$ b) $\dfrac{0.1}{\sqrt{2\pi}}\,e^{-\frac{1}{2}\left(\frac{x-60}{10}\right)^2}$ c) $\dfrac{0.1}{\sqrt{2\pi}}\,e^{-\frac{1}{2}\left(\frac{x-80}{10}\right)^2}$

2.

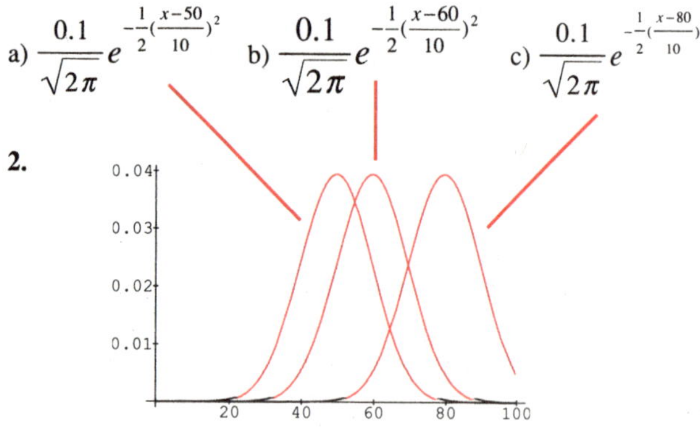

We observe that if μ is increased by k units, the corresponding graph moves k units to the right if $k > 0$ and to the left if $k < 0$. The shape of the graph is unchanged.

3. The equations are: $y =$

a) $\dfrac{0.2}{\sqrt{2\pi}}\,e^{-\frac{1}{2}\left(\frac{x-50}{5}\right)^2}$ b) $\dfrac{0.1}{\sqrt{2\pi}}\,e^{-\frac{1}{2}\left(\frac{x-50}{10}\right)^2}$ c) $\dfrac{0.05}{\sqrt{2\pi}}\,e^{-\frac{1}{2}\left(\frac{x-50}{20}\right)^2}$

4.

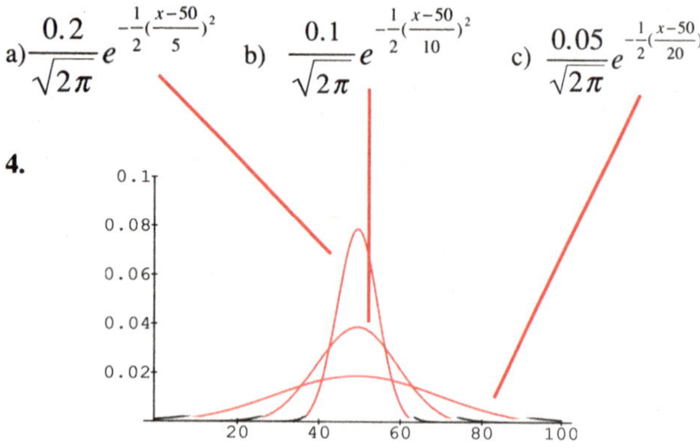

We observe that as σ is increased, by k units, the height of the curve is increased by a factor of k and the graph becomes narrower, with most of its area concentrated close to the mean.

5. a) If the measurement errors are random, i.e. if it is equally likely that a measurement will be larger or smaller than the true value, then the measurements should be roughly equally distributed on both sides of the true value. Since the mean should be close to the true value, then the data should be equally distributed about the mean. That is, the data should by symmetric about the mean.

    b) The values $x = \mu + \varepsilon$ and $x = \mu - \varepsilon$ are equidistant from the line $x = \mu$ and on opposite sides of it (for $\varepsilon > 0$). Substituting both values into the normal equation, we discover that they yield the same value of y.

6. a) and b)

c) We trace as in the worked example from (100,0.026) to (115.06, 0.0168...) to get $\approx 65\%$.

d) Marks of 83 and 117.

$$y = \frac{1}{15\sqrt{2\pi}}\,e^{-\frac{1}{2}\left(\frac{x-100}{15}\right)^2}$$

7. c) Use the range settings: $120 \le x \le 200$; $0 \le y \le 0.08$
Trace from (160.026, 0.0602) to (150.315, 0.0088): about 29%
d) Use similar triangles to find the 95th percentile is about 172 cm

8. a & b)
Using our estimates of m and s, we obtained the equation of the probability function for the SAT scores shown in the display below. When we graphed this equation and the corresponding xyLine with window variables as shown, we obtained the display shown here.

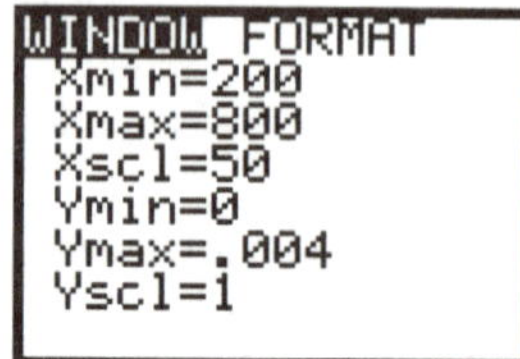

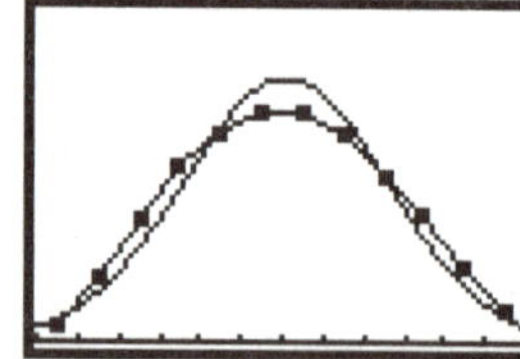

d) about 68%

e) To find the 95th percentile, we can integrate to find the area under the curve up to any value of d using the CALC menu (not the STATS CALC menu, but the one in the top row of keys..)
We find the 95th percentile is close to $x \approx 698$.